Praise for *Fa*

"*Facing the Anthropocene* provides a clear and thorough analysis of how fossil-fuel based capitalism has enabled humans to become a force of nature and how radical political and economic change is our only hope in limiting a far more hostile future. Highly recommended for all those interested in understanding the origins and future of the age of humans."
—CHRISTOPHER WRIGHT, Professor of Organizational Studies, University of Sydney; co-author of *Climate Change, Capitalism and Corporations*

"In this crucial political intervention, Ian Angus combines the work of natural scientists with socialist political economy to produce a synthesis far greater than the sum of its parts, one that can give us the weapons we need. *Facing the Anthropocene* makes quite clear what is to be done."
—MICHAEL LEBOWITZ, author of *The Socialist Imperative: From Gotha to Now*

"With great clarity and unambiguous prose Ian Angus explains why earth has entered a new and dangerous epoch. He embeds the science within its political and economic context, identifying the operation of capitalism as the root cause of our ecological and social planetary emergency. Essential reading for all those who wish to understand the world in order to change it."
—CHRIS WILLIAMS, author of *Ecology and Socialism*

"Why are the earth's ecosystems deteriorating at an ever more rapid pace? What is capitalism's role in creating this new geologic era so influenced by human activity that scientists are calling it the Anthropocene? And what can be done to counter the ominous environmental trends? In *Facing the Anthropocene*, Ian Angus answers these and other questions critical to humanity and the rest of the natural world in a thoughtful, informative, straight-forward, and accessible manner."
—FRED MAGDOFF, co-author of *What Every Environmentalist Needs to Know About Capitalism*

"Using the latest scientific studies, Ian Angus demonstrates the scale of the environmental crises we face and shows how the priorities of the most powerful economies after the Second World War, caused an acceleration of environmental destruction. If we are to survive the Anthropocene, we will need a revolutionary alternative to capitalism. This is a handbook for every activist who wants to be part of that struggle."

—MARTIN EMPSON, author of *Land and Labour: Marxism, Ecology and Human History*

"We face a planetary emergency of which climate warming is the foremost but far from the only manifestation. Ian Angus makes it clear that chaos will not be avoided by gently nudging the prevailing socio-economic system, but only by a global mobilisation for a habitable planet and a new political order."

—DAVID SPRATT, co-author of *Climate Code Red: The Case for Emergency Action*

"*Facing the Anthropocene* traces the biophysical and social roots of the environmental crisis in an attempt to help bridge the divide between the natural and social sciences. This is an important project, and there are not many better suited than Angus to propel it forward."

—UMAIR MUHAMMAD, author of *Confronting Injustice: Social Activism in the Age of Individualism*

"Ian Angus combines forensic judgement with clear words and focused political commitment. He shows that we must understand both the science of climate change and the science of social change. Everyone committed to a sustainable, socially just future should study this book."

—DEREK WALL, International Coordinator, Green Party of England and Wales; author of *Economics after Capitalism: A Guide to the Ruins and a Road to the Future*

"An outstanding contribution, not only for understanding the nature of the Anthropocene and its deadly consequences for human life, but also for explaining its social and economic causes. Ian Angus shows that the catastrophe is not inevitable: there is a possible alternative, based on values of human solidarity. Indispensable reading for ecologists, socialists, climate change activists and rational human beings in general!"

—MICHAEL LÖWY, emeritus research director, National Center for Scientific Research, Paris; author of *Ecosocialism: A Radical Alternative to Capitalist Catastrophe*

Facing the Anthropocene

*Fossil Capitalism and the Crisis
of the Earth System*

by IAN ANGUS

MONTHLY REVIEW PRESS
New York

Library of Congress Cataloging-in-Publication Data:
Names: Angus, Ian, 1945– author.
Title: Facing the anthropocene : fossil capitalism and the crisis of the
  earth system / by Ian Angus.
Description: New York : Monthly Review Press, [2016] | Includes
  bibliographical references and index.
Identifiers: LCCN 2016014078| ISBN 9781583676097 (pbk.) | ISBN 9781583676103
  (hardcover)
Subjects: LCSH: Nature—Effect of human beings on. |
  Capitalism—Environmental aspects. | Global environmental
change—Economic aspects. | Global environmental change—Social aspects.
Classification: LCC GF75 .A64 2016 | DDC 304.2—dc23 LC record available at
https://lccn.loc.gov/2016014078

Typeset in Minion Pro

Monthly Review Press
146 West 29th Street, Suite 6W
New York, New York 10001

www.monthlyreview.org

5 4

Contents

To Lis
my partner in life, love, and hope.
You make everything possible.

it's 3:23 in the morning
and I'm awake
because my great great grandchildren
won't let me sleep
my great great grandchildren
ask me in dreams
what did you do while the planet was plundered?
what did you do when the earth was unraveling?

surely you did something
when the seasons started failing?

as the mammals, reptiles, birds were all dying?

did you fill the streets with protest
when democracy was stolen?

what did you do
once
you
knew?

—From *hieroglyphic stairway*
by Drew Dellinger

Foreword

John Bellamy Foster

For it is because we are kept in the dark about the nature of human society—as opposed to nature in general—that we are now faced (so the scientists concerned assure me), by the complete destructibility of this planet that has barely been made fit to live in.

—BERTOLT BRECHT[1]

The Anthropocene, viewed as a new geological epoch displacing the Holocene epoch of the last 10,000 to 12,000 years, represents what has been called an "anthropogenic rift" in the history of the planet.[2] Formally introduced into the contemporary scientific and environmental discussion by climatologist Paul Crutzen in 2000, it stands for the notion that human beings have become the primary emergent geological force affecting the future of the Earth System. Although often traced to the Industrial Revolution in the late eighteenth century, the Anthropocene is probably best seen as arising in the late 1940s and early 1950s. Recent scientific evidence suggests that the period from around 1950 on exhibits a major spike, marking a Great Acceleration in human impacts on the environment, with the most dramatic stratigraphic trace of the anthropogenic rift to be found in fallout radionuclides from nuclear weapons testing.[3]

Viewed in this way, the Anthropocene can be seen as corresponding roughly to the rise of the modern environmental movement, which had its beginnings in the protests led by scientists against above ground nuclear testing after the Second World War, and was to emerge as a wider movement following the publication of Rachel Carson's *Silent Spring* in 1962. Carson's book was soon followed in the 1960s by the very first warnings, by Soviet and U.S. scientists, of accelerated and irreversible global warming.[4] It is this dialectical interrelation between the acceleration into the Anthropocene and the acceleration of a radical environmentalist imperative in response that constitutes the central theme of Ian Angus's marvelous new book. It is his ability to give us insights into the Anthropocene as a new emergent level of society-nature interaction brought on by historical change—and how the new ecological imperatives it generates have become the central question confronting us in the twenty-first century—that makes *Facing the Anthropocene* so indispensable.

Today it seems likely that the Anthropocene will come to be linked within science to the post–Second World War era in particular. Nonetheless, as in the case of all major turning points in history, there were signs of minor spikes at earlier stages along the way, going back to the Industrial Revolution. This reflects what the Marxian philosopher István Mészáros calls the "dialectic of *continuity and discontinuity*," characterizing all novel emergent developments in history.[5] Although the Anthropocene concept arose fully only with the modern scientific conception of the Earth System, and is now increasingly seen as having its physical basis in the Great Acceleration after the Second World War, it was prefigured by earlier notions, arising from thinkers focusing on the dramatic changes in the human-environmental interface brought on by the rise of capitalism, including the Industrial Revolution, the colonization of the world, and the era of fossil fuels.

"Nature, the nature that preceded human history," Karl Marx and Frederick Engels remarked as early as 1845, "no longer exists anywhere (except perhaps on a few Australian coral islands of recent origin)."[6] Similar views were presented by George Perkins Marsh, in *Man and Nature* in 1864, two years before Ernst Haeckel coined the word *ecology*, and three years before Marx published the first volume of *Capital*, with its warning of the metabolic rift in the human relation to the earth.[7]

It was not until the last quarter of the nineteenth and the early twentieth century, however, that the key concept of the biosphere, out of which our modern notion of the Earth System was to develop, arose, with the publication, most notably, of *The Biosphere* by the Soviet geochemist Vladimir I. Vernadsky in 1926. "Remarkably," Lynn Margulis and Dorian Sagan write in *What Is Life?*, "Vernadsky dismantled the rigid boundary between living organisms and a nonliving environment, depicting life globally before a single satellite had returned photographs of Earth from orbit."[8]

The appearance of Vernadsky's book corresponded to the first introduction of the term Anthropocene (together with Anthropogene) by his colleague, the Soviet geologist Aleksei Pavlov, who used it to refer to a new geological period in which humanity was the main driver of planetary geological change. As Vernadsky oberved in 1945, "Proceeding from the notion of the geological role of man, the geologist A. P. Pavlov [1854–1929] in the last years of his life used to speak of the *anthropogenic era*, in which we now live. . . . He rightfully emphasized that man, under our very eyes, is becoming a mighty and ever-growing geological force. . . . In the 20th Century, man, for the first time in the history of the Earth, knew and embraced the whole biosphere, completed the geographic map of the planet Earth, and colonized its whole surface."[9]

Simultaneously with Vernadsky's work on the biosphere, the Soviet biochemist Alexander I. Oparin and the British socialist biologist J. B. S. Haldane independently developed in the 1920s the theory of the origin of life, known as the "primordial soup theory." As summed up by Harvard biologists Richard Levins and Richard Lewontin, "Life originally arose from inanimate matter [what Haldane famously described as a 'hot dilute soup'], but that origination made its continued occurrence impossible, because living organisms consume the complex organic molecules needed to recreate life *de novo*. Moreover, the reducing atmosphere [lacking free oxygen] that existed before the beginning of life has been converted, by living organisms themselves, to one that is rich in reactive oxygen." In this way, the Oparin-Haldane theory explained for the first time how life could have originated out of inorganic matter, and why the process could not be repeated. Equally significant, life, arising in this way billions of years ago, could be seen as the creator of the biosphere within a complex process of coevolution.[10]

It was Rachel Carson, in her landmark 1963 speech, "Our Polluted Environment," famously introducing the concept of ecosystem to the U.S. public, who most eloquently conveyed this integrated ecological perspective, and the need to take it into account in all of our actions. "Since the beginning of biological time," she wrote,

> there has been the closest possible interdependence between the physical environment and the life it sustains. The conditions on the young earth produced life; life then at once modified the conditions of the earth, so that this single extraordinary act of spontaneous generation could not be repeated. In one form or another, action and interaction between life and its surroundings have been going on ever since.
>
> This historic fact has, I think, more than academic significance. Once we accept it we see why we cannot with impunity make repeated assaults upon the environment as we now do. The serious student of earth history knows that neither life nor the physical world that supports it exists in little isolated compartments. On the contrary, he recognizes the extraordinary unity between organisms and the environment. For this reason he knows that harmful substances released into the environment return in time to create problems for mankind.
>
> The branch of science that deals with these interrelations is Ecology. . . . We cannot think of the living organism alone; nor

can we think of the physical environment as a separate entity. The two exist together, each acting on the other to form an ecological complex or ecosystem.[11]

Nevertheless, despite the integrated ecological vision presented by figures like Carson, Vernadsky's concepts of the biosphere and biogeochemical cycles were for a long time downplayed in the West due to the reductionist mode that prevailed in Western science and the Soviet background of these concepts. Soviet scientific works were well known to scientists in the West and were frequently translated in the Cold War years by scientific presses and even by the U.S. government—though unaccountably Vernadsky's *The Biosphere* was not translated into English until 1998. This was a necessity since in some fields, such as climatology, Soviet scientists were well ahead of their U.S. counterparts. Yet this wider scientific interchange, crossing the Cold War divide, was seldom conveyed to the public at large, where knowledge of Soviet achievements in these areas was practically nonexistent. Ideologically, therefore, the concept of the biosphere seems to have long fallen under a kind of interdict.

Still, the biosphere took center stage in 1970, with a special issue of *Scientific American* on the topic.[12] In that same year the socialist biologist Barry Commoner warned in *The Closing Circle* of the vast changes in the human relation to the planet, beginning with the atomic age and the rise of modern developments in synthetic chemistry. Commoner pointed back to the early warning of capitalism's environmental disruption of the cycles of life represented by Marx's discussion of the rift in the metabolism of the soil.[13]

Two years later, Evgeni K. Fedorov, one of the world's top climatologists and a member of the Presidium of the Supreme Soviet of the USSR, as well as the leading Soviet supporter of Commoner's analysis (writing the "Concluding Remarks" to the Russian edition), declared that the world would need to wean itself from fossil fuels: "A rise in temperature of the earth is inevitable if we do not confine ourselves to the use, as energy sources, of direct solar radiation and the hydraulic energy of wave and wind energy, but [choose instead to] obtain energy from fossil [fuels] or nuclear reactions."[14] For Fedorov, Marx's theory of "metabolism between people and nature" constituted the methodological basis for an ecological approach to the question of the Earth System.[15] It was in the 1960s and 1970s that climatologists in the USSR and the United States first found "evidence," in the words of Clive Hamilton and Jaques Grinevald, of a "worldwide metabolism."[16]

The rise of Earth System analysis in the succeeding decades was also strongly impacted by the remarkable view from outside, emanating from the early space missions. As Howard Odum, one of the leading figures in the formation of systems ecology, wrote in *Environment, Power and Society*:

> We can begin a systems view of the earth through the macroscope of the astronaut high above the earth. From an orbiting satellite, the earth's living zone appears to be very simple. The thin water- and air-bathed shell covering the earth—the biosphere—is bounded on the inside by dense solids and on the outside by the near vacuum of outer space.... From the heavens it is easy to talk of gaseous balances, energy budgets per million years, and the magnificent simplicity of the overall metabolism of the earth's thin outer shell. With the exception of energy flow, the geobiosphere for the most part is a closed system of the type whose materials are cycled and reused.[17]

"The mechanism of overgrowth," threatening this "overall metabolism," Odum went on to state, "is capitalism."[18] Today's concept of the Anthropocene thus reflects, on the one hand, a growing recognition of the rapidly accelerating role of anthropogenic drivers in disrupting the biogeochemical processes and planetary boundaries of the Earth System and, on the other, a dire warning that the world, under "business as usual," is being catapulted into a new ecological phase—one less conducive to maintaining biological diversity and a stable human civilization.

It is the bringing together of these two aspects of the Anthropocene—variously viewed as the geological and the historical, the natural and social, the climate and capitalism—in one single, integrated view, that constitutes the main achievement of *Facing the Anthropocene*. Angus demonstrates that "fossil capitalism," if not stopped, is a runaway train, leading to global environmental apartheid and what the great British Marxist historian E. P. Thompson referred to as the threatened historical stage of "exterminism," in which the conditions of existence of hundreds of millions, perhaps billions of people will be upended, and the very basis of life as we know it endangered. Moreover, this has its source in what Odum called "imperial capitalism," imperiling the lives of the most vulnerable populations on the planet in a system of forced global inequality.[19]

Such are the dangers that only a new, radical approach to social science (and thus to society itself), Angus informs us—one that takes seriously Carson's warning that if we undermine the living processes of Earth this will "return in time" to haunt us—can provide us with the answers that we need in the Anthropocene epoch. Where such urgent change is concerned "tomorrow is too late."[20]

Yet the dominant social science, which serves the dominant social order and its ruling strata, has thus far served to obscure these issues, putting its weight behind ameliorative measures together with mechanistic solutions such as carbon markets and geoengineering—as if the answer to the Anthropocene crisis were a narrowly economic and technological one consistent with the further expansion of the hegemony of capital over Earth and its inhabitants; this despite the fact that the present system of capital accumulation is at the root of the crisis. The result is to propel the world into still greater danger. What is needed, then, is to recognize that it is the logic of our current mode of production—capitalism—that stands in the way of creating a world of sustainable human development transcending the spiraling disaster that otherwise awaits humanity. To save ourselves we must create a different socioeconomic logic pointing to different human-environmental ends: an ecosocialist revolution in which the great mass of humanity takes part.

But are there not risks to such radical change? Would not great struggles and sacrifices attend any attempt to overthrow the prevailing system of production and energy use in response to global warming? Is there any surety that we would be able to create a society of sustainable human development, as ecosocialists like Ian Angus envision? Would it not be better to err on the side of denialism than on the side of catastrophism? Should we not hesitate to take action at this level until we know more?

Here it is useful to quote from the great German playwright and poet Bertolt Brecht's didactic poem, "The Buddha's Parable of the Burning House":

> The Buddha still sat under the bread-fruit tree and to the
>   others,
> To those who had not asked [for guarantees], addressed this
>   parable:
> "Lately I saw a house. It was burning. The flame
> Licked at its roof. I went up close and observed
> That there were people still inside. I entered the doorway and
>   called
> Out to them that the roof was ablaze, so exhorting them
> To leave at once. But those people
> Seemed in no hurry. One of them,
> While the heat was already scorching his eyebrows,
> Asked me what it was like outside, whether it wasn't raining,
> Whether the wind wasn't blowing, perhaps, whether there was
> Another house for them, and more of this kind. Without
>   answering
> I went out again. These people here, I thought,
> Must burn to death before they stop asking questions.
>   And truly, friends,
> Whoever does not yet feel such heat in the floor that he'll gladly

Exchange it for any other, rather than stay, to that man
I have nothing to say." So Gautama the Buddha.[21]

It is capitalism and the alienated global environment it has produced that constitutes our "burning house" today. Mainstream
environmentalists, faced with this monstrous dilemma, have generally chosen to do little more than *contemplate* it, watching and
making minor adjustments to their interior surroundings while
flames lick the roof and the entire structure threatens to collapse
around them. The point, rather, is to *change* it, to rebuild the house
of civilization under different architectural principles, creating
a more sustainable metabolism of humanity and the earth. The
name of the movement to achieve this, rising out of the socialist and radical environmental movements, is *ecosocialism*, and the
book before you is its most up-to-date and eloquent manifesto.

—EUGENE, OREGON
JANUARY 9, 2016

ABBREVIATIONS

AWG	Anthropocene Working Group
BRICS	Brazil, Russia, India, China, South Africa
°C	Celsius degrees
CFC	Chlorofluorocarbon
CH_4	Methane
CIO	Congress of Industrial Organizations
CO_2	Carbon Dioxide
COP	Conference of the Parties (to the UNFCCC)
G20	Group of 20
GDP	Gross Domestic Product
GCF	Green Climate Fund
GM	General Motors
GNP	Gross National Product
ICS	International Commission on Stratigraphy
ICSU	International Council of Scientific Unions
IGBP	International Geosphere-Biosphere Program
IPCC	Intergovernmental Panel on Climate Change
IUGS	International Union of Geological Sciences
MEA	Millennium Ecosystem Assessment
MECW	Marx-Engels Collected Works
NASA	National Aeronautics and Space Administration
NO_x	Nitrogen Dioxide and Nitrous Oxide
O_2	Oxygen
O_3	Ozone
OECD	Organization for Economic Cooperation and Development
PAGES	Past Global Changes project
PIK	Potsdam Institute for Climate Impact Research
ppm	Parts per million
RCP	Representative Concentration Pathway
UN	United Nations
UNFCCC	United Nations Framework Convention on Climate Change
USSR	Union of Soviet Socialist Republics
UV	Ultraviolet
WBGT	Wet Bulb Globe Temperature

Preface

> The earth is polluted neither because man is some kind of especially dirty animal nor because there are too many of us. The fault lies with human society—with the ways in which society has elected to win, distribute, and use the wealth that has been extracted by human labor from the planet's resources. Once the social origins of the crisis become clear, we can begin to design appropriate social actions to resolve it.
>
> —BARRY COMMONER[1]

In the past twenty years, earth science has taken a giant leap forward, combining new research in multiple disciplines to expand our understanding of the Earth System as a whole. A central result of that work has been realization that a new and dangerous stage in planetary evolution has begun—the *Anthropocene*. At the same time, ecosocialists have made huge strides in rediscovering and extending Marx's view that capitalism creates an "irreparable rift in the interdependent process of social metabolism," leading inevitably to ecological crises. These two developments have for the most part occurred separately, and despite their mutual relevance, there has been little interchange between them.

Facing the Anthropocene is a contribution toward bridging the gap between Earth System science and ecosocialism. I hope to show socialists that responding to the Anthropocene must be a central part of our program, theory, and activity in the twenty-first century, and to show Earth System scientists and environmentalists that ecological Marxism provides essential economic and social understanding that is missing in most discussions of the new epoch.

The book's title has two meanings. It refers, first, to the fact that humanity in the twenty-first century faces radical changes in its physical environment—not just more pollution or warmer weather, but a *crisis of the Earth System*, caused by human activity. And it is a challenge to everyone who cares about humanity's future to face up to the fact that survival in the Anthropocene requires radical social change, replacing fossil capitalism with an ecological civilization, ecosocialism.

The global environmental crisis is the most important issue of our time. Fighting to limit the damage caused by capitalism today will help lay the basis for socialism tomorrow, and even then, building socialism in Anthropocene conditions will involve challenges that no twentieth century socialist ever imagined. Understanding and preparing for those challenges must now be at the top of the socialist agenda.

Facing the Anthropocene is not the final word on any of these subjects. I do not have all the answers, and the task before us is immense, so please consider this as the beginning of a discussion, not a final declaration. I look forward to receiving responses, amplifications, and, of course, disagreements. The web journal I edit, *climateandcapitalism.com*, offers a forum for continuing discussion of the issues raised in this book.

The book is divided into three parts.

• *Part One, A No-Analog State.* In the past two decades, little noticed by mainstream media and most environmentalists, scientists have made critically important discoveries about the history and current state of our planet, and have concluded that Earth has entered a new and unprecedented state, an epoch they have named the Anthropocene.

- *Part Two: Fossil Capitalism.* Part One discussed the Anthropocene as a *biophysical* phenomenon, but to properly understand it, we must see it as a *socio-ecological* phenomenon, as a product of the rise of capitalism and its deep dependence on fossil fuels.
- *Part Three: The Alternative.* Another Anthropocene is possible, if the majority of humanity fights back. What should our objectives be, and what kind of movement do we need to achieve them?

Tucked away in the Appendix are two short essays on misunderstandings about the Anthropocene that have some currency on the left: the claim that Anthropocene science blames all humanity for the planetary crisis, and the related assertion that scientists have chosen an inappropriate name for the new epoch.

What This Book Doesn't Do

Debate climate science. The science is unequivocal: greenhouse gas emissions, primarily resulting from burning fossil fuel and deforestation, have significantly increased Earth's average temperature, and continue to do so. The only uncertainties concern how quickly and how high global temperatures will rise if nothing is done to slow or stop emissions. Anyone who denies that is either blind to the science, or deliberately lying: such people are unlikely to read this book, but if they do, they won't be convinced.

Describe fully the planetary emergency. This book is about the discovery, effects, and socioeconomic causes of the Anthropocene, and that emphasis required omitting or limiting discussion of such vital issues as biodiversity loss and freshwater depletion. A large book could be written about each of the nine Planetary Boundaries that are at risk today and still the account would be incomplete. For readers who want to learn more, some suggestions for further reading are posted on *climateandcapitalism.com*.

The Truth Is Always Concrete

Much environmental writing reduces human history to population growth and technology change, both of which just happen
somehow. Why some societies have higher birth rates than others,
why the ancient Greeks only used steam power in toys, why the
Industrial Revolution occurred in England and not India or
China—such questions are not asked. Having defined a set of
abstract ecological principles that apply to all societies at all times,
any further explanation is superfluous.

Socialists are not immune to such reasoning. I have a shelf of
books and pamphlets from various left-wing authors and groups,
all proving that environmental destruction is caused by the accumulation of capital, and all jumping directly from that to a call
for socialism. How are capitalism's anti-ecological characteristics
manifested concretely in the real world? Are today's environmental crises simply new renditions of past problems, or is something
new and different happening? If the latter, how should our strategies and tactics change? All too frequently, such issues are passed
over in silence.

Even more disturbing, in the present context, are articles that
criticize or reject the very concept of the Anthropocene, by left-
wing authors whose first reaction to new science is to warn of
potential political contamination from ideologically suspect
scientists. It seems that for some, anything less than explicit anti-
capitalism must be denounced as a dangerous diversion.

When Charles Darwin published *The Origin of Species* in 1859,
Marx and Engels read it eagerly. They attended public lectures by
prominent scientists whose political views were far from their own.
Their private correspondence shows that they didn't accept every
word Darwin wrote, but they didn't denounce him publicly for
not being a socialist; rather, they did their utmost to incorporate
the latest findings of science into their own work and world-
view. Today's anti-Anthropocene radicals should ask themselves,
"WWMED?—What Would Marx and Engels Do?" What Marx

and Engels *would not* do, we can be sure, is build walls between social and natural science.

Rather than carping from the sidelines about the scientists' lack of social analysis (or worse, rejecting science altogether) socialists need to approach the Anthropocene project as an opportunity to unite an ecological Marxist analysis with the latest scientific research in a new synthesis—a socio-ecological account of the origins, nature, and direction of the current crisis. Moving toward such a synthesis is an essential part of developing a program and strategy for twenty-first-century socialism: if we do not understand what drives capitalism's hell-bound train, we will not be able to stop it.

Nearly fifty years ago, the pioneering environmentalist Barry Commoner warned that "the environmental crisis reveals serious incompatibilities between the private enterprise system and the ecological base on which it depends."[2] It is now time—it is *past* time—to hear his warning and change that system.

Acknowledgments

I owe a particular debt of gratitude to John Bellamy Foster, editor of *Monthly Review* and prolific writer on Marxist ecology and economics. He provided frequent advice and gave me detailed comments and suggestions as my work proceeded, beginning when I proposed a short article on the Anthropocene. This book literally would not have been written without his constant support and encouragement.

Clive Hamilton, Robert Nixon, Peter Sale, Will Steffen, Philip Wright, and Jan Zalasiewicz took time from their work to respond to my emailed questions about their areas of expertise.

Jeff White carefully proofread several drafts, checked the reference notes and identified weaknesses in the text. Lis Angus, John Riddell and Fred Magdoff offered criticisms and insights that helped me to think the subject through and express my ideas more clearly.

The team at Monthly Review Press, Michael Yates, Martin Paddio, and Susie Day, have been a pleasure to work with. Erin Clermont copy-edited my final draft and prepared it for publication.

———————⚓———————

Parts of *Facing the Anthropocene* were previously published in *Climate & Capitalism*, *Monthly Review*, and other publications. All have been rewritten and updated for this book.

Many thanks to Drew Dellinger for permission to include a verse from "hieroglyphic stairway," first published in the collection *Love Letter to the Milky Way* (White Cloud Press, 2011).

"System Change Not Climate Change" in chapter 12 was written by Terry Townsend for *Green Left Weekly* in 2007. Terry, who edits the indispensable *Links Journal of International Socialist Renewal*, kindly granted permission to publish an updated version here.

"The Fertilizer Footprint" in chapter 10 was first published in September 2015 by the nonprofit organization GRAIN, which makes its excellent materials freely available, without copyright.

Third printing, 2018: The description of the Carbon Cycle, on pages 123–4, has been revised.

Metric Measures

All scientific research uses the International System of Units (SI), commonly called the metric system, and this book follows that standard. Temperatures are given in degrees Celsius (°C). One degree Celsius equals 1.8 degrees Fahrenheit, so restricting the global average temperature increase to 2°C means restricting it to 3.6°F. Distances are given in meters and kilometers. One meter equals 3.3 feet. One kilometer equals 0.6 miles. A tonne, sometimes called a metric ton, is 1,000 kilograms, just over 2,200 pounds.

PART ONE

A NO-ANALOG STATE

For nearly a decade after it first appeared in scientific literature in 2000, the word *Anthropocene* remained the exclusive property of specialists in Earth sciences. It was seldom heard, and even less often discussed, outside scientific circles.

But in 2011, a web search for Anthropocene produced over 450,000 hits, "Welcome to the Anthropocene" was a cover headline on *The Economist*, the Royal Society devoted an entire issue of its journal to it, the Dalai Lama held a seminar, and the Vatican commissioned and published a report.

There are now three academic journals devoted to the Anthropocene. It has been the subject of dozens of books, hundreds of academic papers, and innumerable articles in newspapers, magazines, websites, and blogs. There have been exhibitions about art in the Anthropocene, conferences about the humanities in the Anthropocene, novels about love in the Anthropocene, and there's even a heavy metal album called *The Anthropocene Extinction*.

In the comic strip *Dilbert*, when Bob the dinosaur asked his smart watch for the time, the watch replied, "This is the Anthropocene epoch."[1]

Rarely has a scientific term moved so quickly into wide acceptance and general use. Even more rarely has a scientific term been the subject of so much misinformation and confusion. As Australian environmentalist Clive Hamilton has justly complained, much of what is written about the subject appears to have come from "people who have not bothered to read the half-dozen basic papers on the Anthropocene by those who have defined it, and therefore do not know what they are talking about."[2]

This book does not attempt to address all the political and philosophical debates the Anthropocene has generated, nor does it discuss specialized technical questions. It aims, rather, to provide essential background and context for activists who need to understand what the Anthropocene is and why it is important. Such an understanding is essential to the development of an effective ecosocialist movement today, and will be even more critical for building a post-capitalist society tomorrow.

Part One discusses how scientists came to identify a qualitative change in Earth's most critical physical characteristics, and what the implications are for all living things, including humans.

1

A Second Copernican Revolution

In terms of some key environmental parameters, the Earth System has moved well outside the range of the natural variability exhibited over the last half million years at least. The nature of changes now occurring simultaneously in the Earth System, their magnitudes and rates of change, are unprecedented. The Earth is currently operating in a no-analog state.

—AMSTERDAM DECLARATION ON GLOBAL CHANGE[1]

The word *Anthropocene* has been coined three times.

In 1922, the Soviet geologist Aleksei Petrovich Pavlov proposed Anthropocene or Anthropogene as a name for the time since the first humans evolved about 160,000 years ago. Both words were used by Soviet geologists for some time, but they were never accepted in the rest of the world.

In the 1980s, marine biologist Eugene Stoermer used the word in some published articles, but no one seems to have followed his lead.

The third time's the charm. Atmospheric chemist Paul J. Crutzen reinvented the word in February 2000, at a meeting of the International Geosphere-Biosphere Program in Cuernavaca, Mexico. Will Steffen, then executive director of the IGBP, was a witness:

Scientists from IGBP's paleoenvironment project were report-
ing on their latest research, often referring to the Holocene,
the most recent geological epoch of earth history, to set the
context for their work. Paul, a vice-chair of IGBP, was becoming visibly agitated at this usage, and after the term Holocene
was mentioned yet again, he interrupted them: "Stop using the
word Holocene. We're not in the Holocene any more. We're
in the . . . the . . . the . . . (searching for the right word) . . . the
Anthropocene!"[2]

Five years earlier, Crutzen had won a Nobel Prize for work
that helped prove that widely used chemicals were destroying the
ozone layer in Earth's upper atmosphere, with potentially catastrophic effects for all life on Earth. In his acceptance speech, he
said that his research on ozone had convinced him that the balance of forces on Earth had changed dramatically. It was now
"utterly clear," he said, "that human activities had grown so much
that they could compete and interfere with natural processes."[3] His
interjection at the IGBP meeting in 2000 crystallized that insight
in a single word, *Anthropocene*. "I just made up the word on the
spur of the moment," he says. "Everyone was shocked. But it seems
to have stuck."[4]

Crutzen was something of a scientific superstar: according to
the Institute for Scientific Information, between 1991 and 2001 he
was the world's most-cited author in the geosciences.[5] There is no
question that his high profile drew attention to his articles on the
Anthropocene, and eventually helped win broad acceptance for
the idea.

Steffen, Crutzen, and environmental historian John McNeill
subsequently explained the need for a new word this way:

The term Anthropocene . . . suggests that the Earth has now
left its natural geological epoch, the present interglacial state
called the Holocene. Human activities have become so pervasive and profound that they rival the great forces of Nature and

are pushing the Earth into planetary *terra incognita*. The Earth is rapidly moving into a less biologically diverse, less forested, much warmer, and probably wetter and stormier state. [6]

"A no-analog state," "planetary terra incognita"—these phrases are not used lightly. Earth has entered a new epoch, one that is likely to continue changing in unpredictable and dangerous ways. That's not an exaggeration or a guess: it's the central conclusion of one of the largest scientific projects ever undertaken, one that requires us to think about our planet in an entirely new way.

Earth as an Integrated System

Though it has gone unnoticed by most people and unmentioned in mainstream media, scientific understanding of our planet has radically changed in the past three decades. Scientists have long studied various aspects of Earth, using the methods of geology, biology, ecology, physics, and other disciplines. Now many are studying Earth as an *integrated planetary system*—and discovering that human activity is rapidly changing that system in fundamental ways:

> Crucial to the emergence of this perspective has been the dawning awareness of two fundamental aspects of the nature of the planet. The first is that the Earth itself is a single system, within which the biosphere is an active, essential component. In terms of a sporting analogy, life is a player, not a spectator. Second, human activities are now so pervasive and profound in their consequences that they affect the Earth at a global scale in complex, interactive, and accelerating ways; humans now have the capacity to alter the Earth System in ways that threaten the very processes and components, both biotic and abiotic, upon which humans depend.[7]

Studying Earth as a system became possible and necessary in the 1980s. It became possible when new scientific instruments became available—in particular, satellites designed to gather data about the state of the entire Earth and computer systems capable of collecting, transmitting, and analyzing vast quantities of scientific data. It became necessary when scientists and others realized that nuclear weapons, ozone-destroying chemicals, and greenhouse gases could radically remake the world: human activity was causing not just change but *global change*, with potentially disastrous consequences.

Following discussion of global change at meetings of the International Council of Scientific Unions (ICSU) in Warsaw in 1983 and Ottawa in 1985, a series of international symposia and reports recommended creation of a coordinated international research program on global change. As a member of the American Geographical Union wrote, the need went beyond scientific curiosity:

> It was noted that stresses on the support systems that sustain life were building up at an ever-increasing pace as the result of increases in world population, industrial activity, waste products, pollution, and resource exploitation, as well as because of long-term trends in regional climatic change. To preserve or expand the life-support systems during the 21st century, governments of all nations would have to design long-term plans that, while addressing their own specific national goals, would have to be based on basic scientific knowledge of the global terrestrial environment and on anticipated natural and anthropogenic change. The required detailed and quantitative scientific knowledge simply does not yet exist.[8]

In 1986, the ICSU initiated the International Geosphere-Biosphere Program, "the largest, most complex, and most ambitious program of international scientific cooperation ever to be organized."[9] The IGBP's objective was to "describe and understand the interactive physical, chemical, and biological processes that regulate the total Earth system, the unique environment it provides for life, the changes that are occurring in that system, and the manner in which these changes are influenced by human actions."[10]

A secretariat was established in Stockholm in 1988, and some 500 scientists worldwide began planning initial projects. By the early 1990s, the IGBP was coordinating the work of thousands of scientists studying the *Earth System*, a term that has been well defined by Frank Oldfield and Will Steffen:

> In the context of global change, the Earth System has come to mean the suite of interacting physical, chemical, and biological global-scale cycles (often called biogeochemical cycles) and energy fluxes which provide the conditions necessary for life on the planet. More specifically, this definition of the Earth System has the following features:
> - It deals with a materially closed system that has a primary external energy source, the sun.
> - The major dynamic components of the Earth System are a suite of interlinked physical, chemical, and biological processes that cycle (transport and transform) materials and energy in complex, dynamic ways within the System. The forcings and feedbacks within the System are at least as important to the functioning of the System as are the external drivers.
> - Biological/ecological processes are an integral part of the functioning of the Earth System, and not just the recipients of changes in the dynamics of a physico-chemical system. Living organisms are active participants, not simply passive respondents.
> - Human beings, their societies and their activities are an integral component of the Earth System, and are not an outside force perturbing an otherwise natural system. There are many modes of natural variability and instabilities within the System as well as anthropogenically driven changes. By definition, both types of variability are part of

the dynamics of the Earth System. They are often impossible to separate completely and they interact in complex and sometimes mutually reinforcing ways.[11]

As Hans Schellnhuber of the Potsdam Institute for Climate Impact Research wrote, this was a revolutionary shift in the scientific view of Earth, comparable to the sixteenth-century discovery by Copernicus that Earth orbits the Sun.

Optical magnification instruments once brought about the Copernican revolution that put the Earth in its correct astrophysical context. Sophisticated information-compression techniques including simulation modeling are now ushering in a second "Copernican" revolution. . . .

This new revolution will be in a way a reversal of the first: it will enable us to look back on our planet to perceive one single, complex, dissipative, dynamic entity, far from thermodynamic equilibrium—the "Earth system."[12]

Global Change and the Earth System

An overarching goal of the IGBP's work was to develop "a substantive science of integration, putting the pieces together in innovative and incisive ways toward the goal of understanding the dynamics of the planetary life support system as a whole." By early in the twenty-first century, they were confident that "an integrative Earth System science is already beginning to unfold."[13]

In 2000, the IGBP was a decade old, and its various projects had begun preparing comprehensive reports on what had been learned in ten years of Earth System research. The extensive documents that resulted were subsequently published by the German publishing house Springer Verlag, as the IGBP Book Series.[14]

The meeting in Mexico in February 2000 was part of the summing-up process. Paul Crutzen's outburst—"We're in the Anthropocene!"—led to intense unscheduled discussions. For ten years, the participants had been immersed in detailed investigation of aspects of the Earth System; now they saw a theme that unified their work: the Earth System *as a whole* was being qualitatively transformed by human action. That realization confirmed the need for an overall synthesis of scientific knowledge about the past, present, and probable future of the Earth System:

> The synthesis aimed to pull together a decade of research in IGBP's core projects, and, importantly, generate a better understanding of the structure and functioning of the Earth System as a whole, more than just a description of the various parts of the Earth System around which IGBP's core projects were structured. The increasing human pressure on the Earth System was a key component of the synthesis.[15]

Crutzen's proposal crystallized a new perspective on the impact of global change. According to Steffen, "The concept of the Anthropocene became rapidly and widely used throughout the IGBP as its projects pulled together their main findings. The Anthropocene thus became a powerful concept for framing the ultimate significance of global change."[16]

After the February 2000 IGBP meeting, a literature search found that Eugene Stoermer had previously used the word, so Crutzen invited him to co-sign a short article in the IGBP's *Global Change Newsletter*:

> Considering these and many other major and still growing impacts of human activities on earth and atmosphere, and at all, including global, scales, it seems to us more than appropriate to emphasize the central role of mankind in geology and ecology by proposing to use the term "anthropocene" for the current geological epoch.[17]

The article alerted scientists associated with the IGBP that a new synthesizing framework was emerging. A few months later, the message was reinforced by a peer-reviewed article in the prestigious journal *Science*, in which the members of the IGBP's Carbon Cycle Working Group referred to humanity "rapidly enter[ing] a new Earth System domain, the 'Anthropocene' Era."[18]

But the Anthropocene's real coming-out party was in Amsterdam, in July 2001. "Challenges of a Changing Earth," a conference organized jointly by the IGBP, the International Human Dimensions Program, the World Climate Research Program, and the biodiversity program DIVERSITAS, was a critical turning point in the development of Earth System science. About 1,400 people, including researchers from 105 countries, took part in four days of lectures and discussions, many of them focused on the IGBP's research.

The materials that participants were given included a 32-page pamphlet, signed by all four sponsors but obviously prepared by the IPGB. Titled *Global Change and the Earth System: A Planet Under Pressure*, it would later be expanded, using the same title, into the IGBP's 350-page synthesis report. The pamphlet, which is in effect a high-level outline of the later book, included a chapter, "The Anthropocene Era," that expanded on the arguments presented in the Crutzen-Stoermer newsletter article:

> Until very recently in the history of Earth, humans and their activities have been an insignificant force in the dynamics of the Earth System. Today, humankind has begun to match and even exceed nature in terms of changing the biosphere and impacting other facets of Earth System functioning. The magnitude, spatial scale, and pace of human-induced change are unprecedented. Human activity now equals or surpasses nature in several biogeochemical cycles. The spatial reach of the impacts is global, either through the flows of the Earth's cycles or the cumulative changes in its states. The speed of these changes is on the order of decades to centuries, not the

centuries to millennia pace of comparable change in the natural dynamics of the Earth System.

The extent to which human activities are influencing or even dominating many aspects of Earth's environment and its functioning has led to suggestions that another geological epoch, the *Anthropocene Era . . .* has begun:

- in a few generations humankind is in the process of exhausting fossil fuel reserves that were generated over several hundred million years,
- nearly 50% of the land surface has been transformed by direct human action, with significant consequences for biodiversity, nutrient cycling, soil structure and biology, and climate,
- more nitrogen is now fixed synthetically and applied as fertilizers in agriculture than is fixed naturally in all terrestrial ecosystems,
- more than half of all accessible freshwater is used directly or indirectly by humankind, and underground water resources are being depleted rapidly in many areas,
- the concentrations of several climatically important greenhouse gases, in addition to CO_2 and CH_4, have substantially increased in the atmosphere,
- coastal and marine habitats are being dramatically altered; 50% of mangroves have been removed and wetlands have shrunk by one-half,
- About 22% of recognized marine fisheries are overexploited or already depleted, and 44% more are at their limit of exploitation,
- Extinction rates are increasing sharply in marine and terrestrial ecosystems around the world; the Earth is now in the midst of its first great extinction event caused by the activities of a single biological species (humankind).[19]

The pamphlet presented the Anthropocene concept tentatively—it said the changes have "led to suggestions," not that a new geological period had definitely begun. This likely reflected unwillingness by the other three sponsors to endorse a concept that was new to them.

This caution extended to the *Declaration on Global Change* the conference adopted. Although it said that "the Earth System has moved well outside the range of the natural variability exhibited over the last half million years at least," and that "Earth is currently operating in a no-analog state," the Declaration did not mention a new geological epoch or use the word *Anthropocene*.[20]

After the Amsterdam conference, Paul Crutzen submitted a more strongly worded article to *Nature*, one of the world's most widely read scientific journals. The oddly titled "Geology of Mankind," published in January 2002, was the first peer-reviewed paper to specifically argue that a new geological epoch had begun.

Again Crutzen listed ways in which human activity was changing the face of Earth, including:

- A tenfold human population growth in three centuries.
- Maintaining 1.4 billion methane-producing cattle.
- Exploiting 20–50 percent of Earth's land surface.
- Destruction of tropical rainforests.
- Widespread dam building and river diversion.
- Exploitation of more than half of all accessible fresh water.
- A 25 percent decline of fish in upwelling ocean regions and 35 percent in the continental shelf.
- A 16-fold increase in energy use in the twentieth century, raising sulphur dioxide emissions to over twice natural levels.
- Use of more than twice as much nitrogen fertilizer in agriculture as is used naturally in all terrestrial ecosystems combined.
- Increasing atmospheric concentrations of greenhouse gases to their highest levels in over 400,000 years.

He pointed to global consequences, including acid precipitation, photochemical smog and global warming of 1.4 to 5.8 degrees Celsius during this century. He was careful to add that "these effects have largely been caused by only 25% of the world population."

Barring a global catastrophe such as a meteorite impact, world war, or pandemic, Crutzen wrote, "mankind will remain a major environmental force for many millennia," and so "it seems appropriate to assign the term 'Anthropocene' to the present, in many ways human-dominated, geological epoch."[21]

A New Synthesis

Meanwhile, an eleven-person team headed by Will Steffen had begun the complex and time-consuming task of synthesizing a decade's work by thousands of scientists into a single volume that would be largely accessible to a non-expert audience. Steffen says the main text was "a true synthesis, as we did not assign chapters to individual authors but rather wrote the whole book as a single, integrated narrative with all authors contributing to the whole book."[22] The team also commissioned and included short essays by individual experts to highlight important aspects of the subject.

Completed early in 2003 and published in 2004, *Global Change and the Earth System: A Planet Under Pressure* (not to be confused with the earlier pamphlet of the same name) was an invaluable contribution to broad understanding of the Earth System—and despite the many scientific advances that have been made since, it remains essential reading for anyone who wants to understand the scientific basis for declaring a new epoch, the Anthropocene.[23]

2

The Great Acceleration

We know that *something* went wrong in the country after World War II, for most of our serious pollution problems either began in the postwar years or have greatly worsened since then.

—BARRY COMMONER[1]

At some point the IGBP team that prepared *Global Change and the Earth System* decided that their book should "record the trajectory of the 'human enterprise' through a number of indicators" from 1750 to 2000.[2] The result was 24 graphs—twelve showing historical trends in human activity (GDP growth, population, energy consumption, water use, etc.) and twelve showing physical changes in the Earth System (atmospheric carbon dioxide, ozone depletion, species extinctions, loss of forests, etc.) over 250 years.

The authors were surprised by what they found: Every trend line showed gradual growth from 1750 and a sharp upturn from about 1950. "We expected to see a growing imprint of the human enterprise on the Earth System from the start of the Industrial Revolution onward. We didn't, however, expect to see the dramatic change in magnitude and rate of the human imprint from about 1950 onward."[3] They pointed this out in the book:

One feature stands out as remarkable. The second half of the twentieth century is unique in the entire history of human existence on Earth. Many human activities reached take-off points sometime in the twentieth century and have accelerated sharply towards the end of the century. The last 50 years have without doubt seen the most rapid transformation of the human relationship with the natural world in the history of humankind.[4]

Millennium Ecosystem Assessment

While the IGBP was preparing its synthesis report, another global scientific project was completing its work. The Millennium Ecosystem Assessment (MEA), coordinated by the United Nations Environment Program, was launched in 2001 to collect and synthesize "authoritative scientific knowledge concerning the impact of changes to the world's ecosystems on human livelihoods and the environment."[5] Nearly 1,400 scientists from around the world contributed to the seven synthesis reports, four technical volumes, and many supporting papers that the MEA published in 2004 and 2005.

One of the project's most important conclusions was highlighted in a final statement from the MEA Board in March 2005. After noting that human societies have always changed the natural systems of the planets to meet their needs, the Board declared that "throughout human history, no period has experienced interference with the biological machinery of the planet on the scale witnessed in the second half of the twentieth century."[6]

The MEA Synthesis Report on *Ecosystems and Human Well-Being* made the same point, and listed significant examples:

> Over the past 50 years, humans have changed ecosystems more rapidly and extensively than in any comparable period of time in human history, largely to meet rapidly growing demands for food, fresh water, timber, fiber, and fuel. This has resulted in a substantial and largely irreversible loss in the diversity of life on Earth. . . .

- More land was converted to cropland in the 30 years after 1950 than in the 150 years between 1700 and 1850. Cultivated systems (areas where at least 30% of the landscape is in croplands, shifting cultivation, confined livestock production, or freshwater aquaculture) now cover one-quarter of Earth's terrestrial surface.

- Approximately 20% of the world's coral reefs were lost and an additional 20% degraded in the last several decades of the twentieth century, and approximately 35% of mangrove area was lost during this time (in countries for which sufficient data exist, which encompass about half of the area of mangroves).

- The amount of water impounded behind dams quadrupled since 1960, and three to six times as much water is held in reservoirs as in natural rivers. Water withdrawals from rivers and lakes doubled since 1960; most water use (70% worldwide) is for agriculture.

- Since 1960, flows of reactive (biologically available) nitrogen in terrestrial ecosystems have doubled, and flows of phosphorus have tripled. More than half of all the synthetic nitrogen fertilizer, which was first manufactured in 1913, ever used on the planet has been used since 1985.

- Since 1750, the atmospheric concentration of carbon dioxide has increased by about 32% (from about 280 to 376 parts per million in 2003), primarily due to the combustion of fossil fuels and land use changes. Approximately 60% of that increase (60 parts per million) has taken place since 1959.

Humans are fundamentally, and to a significant extent irreversibly, changing the diversity of life on Earth, and most of these changes represent a loss of biodiversity.

- More than two-thirds of the area of 2 of the world's 14 major terrestrial biomes and more than half of the area of 4 other biomes had been converted by 1990, primarily to agriculture.

- Across a range of taxonomic groups, either the population size or range or both of the majority of species is currently declining.
- The distribution of species on Earth is becoming more homogenous; in other words, the set of species in any one region of the world is becoming more similar to the set in other regions primarily as a result of introductions of species, both intentionally and inadvertently in association with increased travel and shipping.
- The number of species on the planet is declining. Over the past few hundred years, humans have increased the species extinction rate by as much as 1,000 times over background rates typical over the planet's history (medium certainty). Some 10–30% of mammal, bird, and amphibian species are currently threatened with extinction (medium to high certainty). Freshwater ecosystems tend to have the highest proportion of species threatened with extinction.
- Genetic diversity has declined globally, particularly among cultivated species.[7]

Naming the Turning Point

Almost simultaneously, two large-scale global scientific projects—the International Geosphere-Biosphere Program and the Millennium Ecosystem Assessment—independently identified the middle of the twentieth century as a turning point in Earth history. As the IGBP report said, "The last 50 years have without doubt seen the most rapid transformation of the human relationship with the natural world in the history of the species."[8]

In 2005, Will Steffen and Paul Crutzen of the IGBP, together with environmental historian John McNeill and others who had participated in the MEA process, attended an intensive one-week seminar in Dahlem, Germany, with the aim of deepening their understanding of the history of the relationship between humanity and nature. Their workshop, chaired by Steffen, drew on findings from the IGBP and MEA to argue that "the 20th century can be characterized by global change processes of a magnitude which never occurred in human history." After quoting the MEA, their workshop report gave those processes a name:

> These and many other changes demonstrate a distinct increase in the rates of change in many human-environment interactions as a result of amplified human impact on the environment after World War II—a period that we term the "Great Acceleration."[9]

Steffen later wrote that the name Great Acceleration was a deliberate homage to *The Great Transformation*, Karl Polanyi's influential book on the social, economic, and political upheavals that accompanied the rise of market society in England:

> Polanyi put forward a holistic understanding of the nature of modern societies, including mentality, behavior, structure, and more. In a similar vein, the term "Great Acceleration" aims to capture the holistic, comprehensive, and interlinked nature of the post-1950 changes simultaneously sweeping across the socioeconomic and biophysical spheres of the Earth System, encompassing far more than climate change.[10]

A Two-Stage Anthropocene?

The first peer-reviewed account of the Great Acceleration was the provocatively titled article, "The Anthropocene: Are Humans Now Overwhelming the Great Forces of Nature?" by Steffen, Crutzen, and McNeill, published in 2007. They argued that the Anthropocene had developed in two distinct stages.

Stage 1: The Industrial Era, from the early 1800s to 1945, when atmospheric CO_2 exceeded the upper limit of Holocene variation; and *Stage 2*: The Great Acceleration, from 1945 to the present, "when the most rapid and pervasive shift in the human-environment relationship began."

(They also—over-optimistically, I'd say—predicted that a third stage, "Stewards of the Earth," would begin in 2015.)

Steffen, Crutzen, and McNeil left no doubt that their answer to the question in their article's title was an emphatic *yes*:

> Over the past 50 years, humans have changed the world's ecosystems more rapidly and extensively than in any other comparable period in human history. The Earth is in its sixth great extinction event, with rates of species loss growing rapidly for both terrestrial and marine ecosystems. The atmospheric concentrations of several important greenhouse gases have increased substantially, and the Earth is warming rapidly. More nitrogen is now converted from the atmosphere into reactive forms by fertilizer production and fossil fuel combustion than by all of the natural processes in terrestrial ecosystems put together. . . .
>
> The exponential character of the Great Acceleration is obvious from our quantification of the human imprint on the Earth System, using atmospheric CO_2 concentration as the indicator. Although by the Second World War the CO_2 concentration had clearly risen above the upper limit of the Holocene, its growth rate hit a take-off point around 1950. Nearly three-quarters of the anthropogenically driven rise in CO_2 concentration has occurred since 1950 (from about 310 to 380 ppm), and about half of the total rise (48 ppm) has occurred in just the last 30 years.[11]

The term Great Acceleration quickly caught on among Earth System scientists as a descriptive name for the period of unprecedented economic growth and environmental devastation since World War II. Their "two stages" model has not survived, however; as we'll see in chapter 4 many Earth System scientists, including Steffen, Crutzen, and McNeill, have concluded that the Anthropocene actually began in the middle of the twentieth century, that the beginning of the Great Acceleration is also the beginning of the Anthropocene.

FIGURE 2.1: Earth System Trends

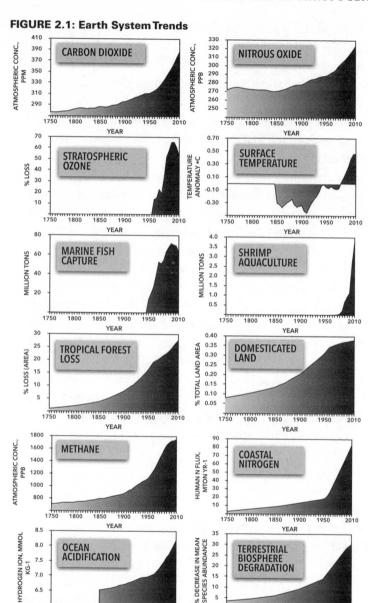

Updates of the 2004 Great Acceleration graphs were published in 2015. As in the original graphs, all the trend lines show hockey stick–shaped trajectories, turning sharply upward in the middle of the twentieth century.[12]

FIGURE 2.2: Socioeconomic Trends

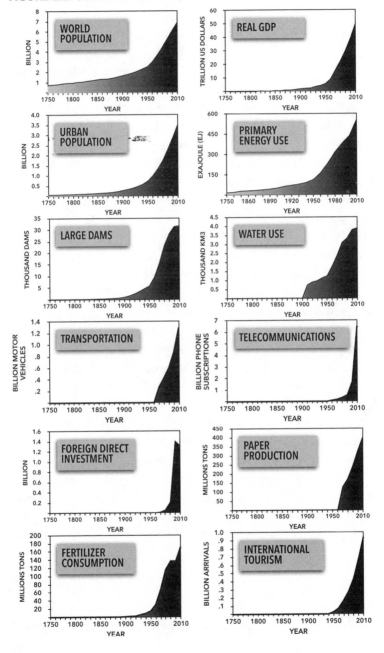

Acceleration Update

The original Great Acceleration graphs, published in 2004, showed social and environmental trends from 1750 to 2000. In January 2015, Will Steffen and others associated with the IGBP and the Stockholm Resilience Center updated the data to 2010, to show what they labelled "The Trajectory of the Anthropocene." In a few cases, where better data was available, they changed the indicators, but on the whole the updated graphs provide a good picture of how the socioeconomic and Earth System trends developed in the first decade of this century.

Overall, both the socioeconomic and Earth System trends show continuing acceleration. The authors note in particular that "the rise in carbon dioxide concentration parallels closely the rise in primary energy use and in GDP, showing no sign yet of any significant decoupling of emissions from either energy use or economic growth."[13]

Two of the Earth System trends do show small declines between 2001 and 2010. One, stratospheric ozone depletion, appears to be the result of an international treaty banning many of the chemicals that are known to destroy ozone. We'll discuss that further in chapter 5.

The downtick in another of the graphs, Marine Fish Capture, is actually bad news for the environment: it reflects the growing exhaustion of the world's ocean fish stocks, leading to a shift from wild fish to farmed fish, which now account for half of global fish consumption.

The amount of domesticated land continues to grow, but unlike the other trends, the rate of growth has been slowing down since 1950. Again, this isn't good news, as it reflects not more careful use of land but a decline in the amount of arable land available. Most of the land now being converted to agriculture was formerly tropical forest, so the indicator for tropical forest loss continues to accelerate.

The Equity Issue

The 2015 update is particularly noteworthy for the authors' thoughtful consideration of the fact that the original graphs displayed global totals, and "did not attempt to deconstruct the socioeconomic graphs into countries or groups of countries." They note that this approach has "prompted some sharp criticism from social scientists and humanities scholars" on the grounds that "strong equity issues are masked by considering global aggregates only."

Steffen and his associates accepted that criticism of the graphs, and went to substantial effort to separate the socioeconomic indicators into three groups: the rich OECD countries, the emerging (BRICS) nations, and the rest of the world. In addition to publishing current versions of the original aggregated graphs, they have added ten graphs that display the socioeconomic indicators for the three groups of countries separately. (There was insufficient data for the other two indicators.)

In a section headed "Deconstructing the Socioeconomic Trends: The Equity Issue," they draw this conclusion: "In 2010 the OECD countries accounted for 74% of global GDP but only 18% of the global population. Insofar as the imprint on the Earth System scales with consumption, most of the human imprint on the Earth System is coming from the OECD world."[14]

In the Appendix of this book I consider the claim made by some on the left that Anthropocene scientists have blamed all of humanity for the actions of a small minority. The 2015 update directly contradicts such charges. Steffen and his associates have clearly shown that they understand the importance of including global inequality as a key factor in any discussion of the causes and effects of the Great Acceleration.

Of course, ecosocialists would take the disaggregation further, breaking out inequalities not just between but within countries, stressing the fact that 1 percent of the population owns half of the world's wealth and that inequality is growing at unprecedented rates. An ecosocialist analysis of the Great Acceleration will build on the decisive issues of class and power that are shaping the Anthropocene and will ultimately determine humanity's future.

3

When did the Anthropocene Begin?

This was a basic tenet of geological science: that human chronologies were insignificant compared with the vastness of geological time; that human activities were insignificant ompared with the force of geological processes. And once they were. But no more.

—NAOMI ORESKES[1]

In 2008, Anthropocene was accurately described as "a vivid yet informal metaphor of global environmental change."[2] Although it was proposed as a new interval of geological time, it had not been defined in geological terms. It is noteworthy, for example, that the scientists who used the word described it variously as a new age, epoch, or era, even though those terms have distinct meanings in geology. Some academic papers treated it as little more than an informal label for the period since the Industrial Revolution, without reference to qualitative changes in the Earth System.[3] None of the principal authors of *Global Change and the Earth System* were geologists, and the IGBP does not seem to have formally or informally submitted the concept to geological organizations for consideration.

As long as it remained informal, Anthropocene was convenient shorthand for a wide variety of phenomena, but its scientific usefulness was limited by the lack of a specific definition based on objective criteria. Loosely defined and even undefined words are widely used in casual conversation, but in science lack of clear definitions can cause confusion.

Fortunately, some geologists set out on their own to determine whether a *prima facie* case could be made for formally defining the Anthropocene as a new geological period, using appropriate geological criteria. And that led to a question that has implications far beyond geology: When did the Anthropocene begin?

The Geological Time Scale

Geologists divide Earth's 4.5 billion–year history into a hierarchy of time intervals—eons, eras, periods, epochs, and ages—called the geological time scale. We live in the Quaternary Period, the most recent subdivision of the Cenozoic Era, which began 65 million years ago. The Quaternary in turn is divided into two epochs—the Pleistocene, which began 2.58 million years ago, and the Holocene, from 11,700 years ago to the present.

The divisions are not arbitrary: they reflect major changes in the dominant conditions and forms of life on Earth, as revealed in geological strata—layers laid down over time in rock, sediment, and ice. The Cenozoic Era is marked by the rise of mammals, following the mass extinction of dinosaurs and most other plants and animals at the end of the Mesozoic. The Pleistocene Epoch was characterized by the repeated expansions and contractions of continental ice sheets in the Northern Hemisphere that are popularly called Ice Ages. The last glacial retreat marks the beginning of the Holocene, which has been characterized by a stable, relatively warm climate: all human history since shortly before the invention of agriculture has occurred in Holocene conditions.

So for geologists, formally approving the Anthropocene is not like applying a faddish label to a current trend, comparable to the Jazz Age or the Gay Nineties. It would mean declaring, on clear scientific criteria, that the present is as different from the Holocene as the Holocene was from the Pleistocene before it.

The subdiscipline of geology that studies and sets standards for geological strata is *stratigraphy*, and it was the Stratigraphic Commission of the Geological Society of London, the world's second-largest organization of geologists, that decided to initiate a review. In late 2007, after a year of investigation, the commission submitted a paper to the journal of the world's largest geological association, the Geological Society of America, which featured it on the cover of the February 2008 issue. The title was a question: "Are We Now Living in the Anthropocene?"[4]

Although "increasing levels of human influence" can be seen in thousands of years of Holocene strata, the authors concluded that before the Industrial Revolution, "human activity did not create new, global environmental conditions that could translate into a fundamentally different stratigraphic signal." Since then, however, "the exploitation of coal, oil, and gas in particular has enabled planet-wide industrialization, construction, and mass transport," producing a wide range of changes that leave traces in strata around the world. The commission focused on four areas of current and expected change that might leave traces for future geologists.

- Increased erosion now exceeds natural sediment production by an order of magnitude.
- Carbon dioxide and methane levels are significantly higher than at any time in nearly a million years, and are rising much faster than in any previous warming period.
- Mass extinctions, species migrations, and replacement of natural vegetation with agricultural monocultures are changing the nature of the biosphere.
- Sea level rises may reach ten to thirty meters for each 1°C increase in temperature, and acidification of ocean water will have severe effects on coral reefs and plankton.

The combined impact of these changes "makes it likely that we have entered a stratigraphic interval without close parallel" in the Quaternary Period, but "it is too early to state whether or not the

Quaternary has come to an end." The authors conservatively concluded that the Anthropocene should be evaluated as a new epoch, not a new period.

> Earth has endured changes sufficient to leave a global stratigraphic signature distinct from that of the Holocene or of previous Pleistocene interglacial phases, encompassing novel biotic, sedimentary, and geochemical change. These changes, although likely only in their initial phases, are sufficiently distinct and robustly established for suggestions of a Holocene-Anthropocene boundary in the recent historical past to be geologically reasonable. . . .
>
> Sufficient evidence has emerged of stratigraphically significant change (both elapsed and imminent) for recognition of the Anthropocene . . . as a new geological epoch to be considered for formalization by international discussion.

This tentative "yes, there is enough evidence for the subject to be considered" received a remarkably quick response. Within months, the International Commission on Stratigraphy (ICS), the committee of the International Union of Geological Sciences (IUGS) that has responsibility for the geological time scale, asked Jan Zalasiewicz, chair of the London Society's Stratigraphic Commission, to convene an international Anthropocene Working Group to investigate and report on whether to formally define the Anthropocene as a geological epoch.

To recommend such a change, the AWG must find that there have been major, qualitative changes to the Earth System, and that geological evidence preserved in rock, sediment, or ice uniquely differentiates layers laid down in the Anthropocene from earlier times. To define when the Holocene/Anthropocene transition occurred, they must propose either a specific stratigraphic marker (often called a "golden spike") or a specific date, or both.[5]

The Anthropocene Working Group includes some 38 volunteer members from 13 countries on five continents. About half are geologists; the rest have backgrounds in other earth sciences, archaeology, and history. They hope to make recommendations during the 35th International Geological Congress in South Africa in August 2016, but formalization of the Anthropocene is not a foregone conclusion. The recommendation might be that the term should remain informal, or that a decision should be delayed. If the AWG recommends formalization, the geological time scale still will not be changed unless 60 percent majorities in the ICS and the IUGS agree.

As paleontologist Anthony Barnosky says, if the Anthropocene gets through all those hoops, "it would not only be a very big deal for earth scientists—the academic equivalent of, say, adding a new amendment to the United States Constitution—but it would also underscore that people have become a geological force every bit as powerful as the kinds of forces that turned an ice-covered Earth into a warm planet, or that wiped out the dinosaurs."[6]

In his first articles on the Anthropocene, Paul Crutzen suggested that the new epoch may have begun at the time of the Industrial Revolution, when large-scale burning of coal launched a long-term rise in atmospheric concentrations of greenhouse gases. That led some observers to conclude that the issue had been prejudged, and many words have been wasted criticizing or praising Crutzen and his co-thinkers for supposedly believing (as some green theorists do) that *industrialization as such* is the source of all environmental problems. Actually, Crutzen was opening a discussion, not declaring a conclusion: he clearly stated that "alternative proposals can be made."[7]

And in fact a dozen or more proposals for dating the Anthropocene have been made to the AWG. Though they differ substantially from one another, the starting dates under serious consideration fall into two broad groups that can be labelled *Early* and *Recent*, depending on whether the proposed starting date is in the distant past, or relatively close to the present.

An Early Anthropocene?

The first Early Anthropocene proposal was advanced by American geologist William Ruddiman, who argues that the Anthropocene began when humans began large-scale agriculture in various parts of the world between eight and five thousand years ago. Those activities, he believes, produced carbon dioxide and methane emissions that raised global temperatures just enough to prevent a return to the Ice Age.[8]

Other Early Anthropocene arguments suggest dating the Anthropocene from the first large-scale landscape modifications by humans, from the extinction of many large mammals in the late Pleistocene, or from the formation of anthropogenic soils in Europe. One widely discussed proposal focuses on the intercontinental exchange of species that followed the European invasions of the Americas, and proposes 1610 as a transition date. Some archaeologists propose to extend the beginning of the Anthropocene back to the earliest surviving traces of human activity, which would take in much of the Pleistocene, and others have suggested that the entire Holocene should simply be renamed Anthropocene, since it is the period when settled human civilizations first developed.

This outpouring of proposals reflects humanity's long and complex relationships with Earth's ecosystems—many of the proposed beginnings are significant turning points in those relationships, and deserve careful study. But the current discussion is not just about human impact: "The Anthropocene is not defined by the broadening impact of humans on the environment, but by active human interference in the processes that govern the geological evolution of the planet."[9] None of the Early Anthropocene options meet that standard, and none of them led to a qualitative break with Holocene conditions.

Even if Ruddiman's controversial claim that the agricultural revolution caused some global warming is correct, that would only mean that human activity had extended Holocene conditions. The recent shift out of Holocene conditions, to a no-analog state, would still need to be evaluated and understood. Noted climatologist James Hansen and his colleagues write:

> Even if the Anthropocene began millennia ago, a fundamentally different phase, a Hyper-Anthropocene, was initiated by explosive 20th-century growth of fossil fuel use. Human-made climate forcings now overwhelm natural forcings. CO_2, at 400 ppm in 2015, is off the scale. . . . Most of the forcing growth occurred in the past several decades, and two-thirds of the 0.9°C global warming (since 1850) has occurred since 1975.[10]

The idea of an Early Anthropocene has been promoted by anti-environmentalist lobbyists associated with the Breakthrough Institute, because it supports their claim that there has been no recent qualitative change and thus there is no need for a radical response. In their view, today's environmental crises "represent an acceleration of trends going back hundreds and even thousands of years earlier, not the starting point of a new epoch."[11]

As Clive Hamilton and Jacques Grinevald explain, the Early Anthropocene argument is attractive to conservatives because it minimizes recent changes to the Earth System:

> It "gradualizes" the new epoch so that it is no longer a rupture due principally to the burning of fossil fuels but a creeping phenomenon due to the incremental spread of human influence over the landscape. This misconstrues the suddenness, severity, duration, and irreversibility of the Anthropocene, leading to a serious underestimation and mischaracterization of the kind of human response necessary to slow its onset and ameliorate its impacts.[12]

A Recent Anthropocene?

The various Early Anthropocene proposals have been considered carefully and rejected by a substantial majority of the Anthropocene

Working Group. In January 2015, over two-thirds of the members signed an article titled "When Did the Anthropocene Begin?: A Mid-Twentieth-Century Boundary Level Is Stratigraphically Optimal."

> Humans started to develop an increasing, but generally regional and highly diachronous, influence on the Earth System thousands of years ago. With the onset of the Industrial Revolution, humankind became a more pronounced geological factor, but in our present view it was from the mid-20th century that the worldwide impact of the accelerating Industrial Revolution became both global and near-synchronous.[13]

They rejected the Early Anthropocene proposals because they only address one aspect of the case for a new epoch, human impact on terrestrial ecosystems. "The significance of the Anthropocene lies not so much in seeing within it the 'first traces of our species' (i.e., an anthropocentric perspective upon geology), but in the scale, significance, and longevity of change (that happens to be currently human-driven) to the Earth System."[14]

In January 2016, the AWG majority published a particularly strong statement on whether changes to the Earth System have been sufficient to justify declaring a new epoch, and if so, when the new epoch began. The title of their paper, published in *Science* magazine and signed by twenty-four AWG members, is unequivocal: "The Anthropocene is functionally and stratigraphically distinct from the Holocene."

Interviewed by *The Guardian*, Colin Waters, lead author of the paper, described the global shift as "a step-change from one world to another that justifies being called an epoch. What this paper does is to say the changes are as big as those that happened at the end of the last ice age. This is a big deal."[15]

The January 2016 paper summarized recent research that identifies major ways in which Holocene conditions no longer exist:

- Atmospheric concentrations of CO_2 have exceeded Holocene levels since at least 1850, and from 1999 to 2010 they rose about 100 times faster than during the increase that ended the last ice age. Methane concentrations have risen further and faster.
- For thousands of years global average temperatures were slowly falling, a result of small cyclical changes in the Earth's orbit. Since 1800, increased greenhouse gases have overwhelmed the orbital climate cycle, causing the planet to warm abnormally rapidly.
- Between 1906 and 2005, the average global temperature increased by up to 0.9°C, and over the past 50 years the rate of change doubled.
- Average global sea levels began rising above Holocene levels between 1905 and 1945. They are now at their highest in about 115,000 years, and the rate of increasing is accelerating.
- Species extinction rates are far above normal. If current trends of habitat loss and overexploitation continue, 75 percent of species could die out in the next few centuries. This would be Earth's sixth mass extinction event, equivalent to the extinction of the dinosaurs, 65 million years ago.

A particularly frightening observation: *even if emission levels are reduced*, by 2070 Earth will be the hottest it has been in 125,000 years, which means it will be "hotter than it has been for most, if not all, of the time since modern humans emerged as a species 200,000 years ago."

Much of the paper focused on a key question for geologists: Has human activity produced a stratigraphic signature in sediments and ice that is distinct from the Holocene? It turns out, contrary to the doubts some expressed at the beginning of this process, that future geologists will have a wealth of indicators to choose from:

> Recent Anthropogenic deposits contain new minerals and rock types, reflecting rapid global dissemination of novel materials including elemental aluminum, concrete, and plastics that

form abundant, rapidly evolving "technofossils." Fossil fuel combustion has disseminated black carbon, inorganic ash spheres, and spherical carbonaceous particles worldwide, with a near-synchronous global increase around 1950.

Anthropocene ice and sediments are also marked by unique concentrations of chemicals, such as lead from gasoline, nitrogen and phosphorus from fertilizers, and carbon dioxide from burning fossil fuels. But "potentially the most widespread and globally synchronous anthropogenic signal is the fallout from nuclear weapons testing." Residues from hydrogen bomb explosions that began in 1952 peaked in 1961–62, leaving a clear worldwide signature.

Each of these stratigraphic signatures is either entirely new or outside of the Holocene range of variability—and the changes are accelerating. The paper recommended that the International Commission on Stratigraphy accept the Anthropocene as a new epoch.

On the question of when the Anthropocene began, the authors' analysis was "more consistent with a beginning in the mid-20th century" than with earlier proposed dates. They did not make a specific midcentury recommendation beyond noting that a number of options have been suggested, ranging from 1945 to 1964.

Finally, they left open the question of "whether it is helpful to formalize the Anthropocene or better to leave it as an informal, albeit solidly founded, geological time term, as the Precambrian and Tertiary currently are."

This is a complex question, in part because, quite unlike other subdivisions of geological time, the implications of formalizing the Anthropocene reach well beyond the geological community. Not only would this represent the first instance of a new epoch having been witnessed firsthand by advanced human societies, it would be one stemming from the consequences of their own doing.

It is still possible that the usually conservative International Commission on Stratigraphy will either reject, or decide to defer, any decision on adding the Anthropocene to the geological time scale, but as the AWG majority writes, "The Anthropocene already has a robust geological basis, is in widespread use, and indeed is becoming a central, integrating concept in the consideration of global change."

In other words, failure to win a formal vote will not make the Anthropocene go away.

4

Tipping Points, Climate Chaos, and Planetary Boundaries

The Anthropocene raises a new question: What are the non-negotiable planetary preconditions that humanity needs to respect in order to avoid the risk of deleterious or even catastrophic environmental change at continental to global scales?

—JOHAN ROCKSTRÖM [1]

After listing recent critical changes to the Earth System, the authors of *Global Change and the Earth System* insisted that such lists do not give the whole picture: "Listing the broad suite of biophysical and socioeconomic changes that is taking place fails to capture the complexity and connectivity of global change since the many linkages and interactions among the individual changes are not included." The listed crises, and others, reinforce and transform one another, producing complex "syndromes of change," and "many changes do not occur in a linear fashion, but rather, thresholds are passed and rapid, non-linear changes ensue.[2]

That understanding of ecological volatility, a recent development in Earth System science, is a direct result of IGBP projects conducted around the world in the 1990s.

The Past as Guide to the Future

From the early 1990s, the International Geosphere-Biosphere Program organized its work into nine projects that focused on broad aspects of the Earth System, including terrestrial ecosystems, atmospheric chemistry, and ocean ecosystems. Each project included a multitude of specific studies conducted by scientists around the world.

All the projects contributed to IGBP's goal of producing an integrated picture of the nature and direction of global change, but arguably the most important, in both objectives and results, was the Past Global Changes (PAGES) project, charged with "providing a quantitative understanding of the Earth's past climate and environment."[3] The importance of this work can be stated simply: we cannot understand the dynamics and direction of today's changing Earth unless we know how current conditions differ from those of the past:

> Understanding the expression, causes, and consequences of past natural variability is of vital concern for developing realistic scenarios of the future. Moreover, the complex interactions between external forcings and internal system dynamics on all timescales implies that at any point in time, the state of the Earth System reflects not only characteristics that are an indication of contemporary processes, but others that are inherited from past influences, all acting on different timescales. The need for an understanding of Earth System functioning that is firmly rooted in knowledge of the past is essential.[4]

To achieve that understanding, researchers needed information about not just a few decades or centuries, but about tens and hundreds of thousands of years for which there are no written or instrumental records. When the IGBP started work, scientists knew some of the deep history of climate in broad outline—when ice ages had occurred, for example—but detail was lacking.

During the 1990s, scientists associated with PAGES conducted unprecedented studies into the physical records that global change leaves behind, including tree rings, coral reefs, ocean and lake sediments, and especially glaciers in which ice has been accumulating in layers for millennia. New methods of extracting and analyzing deep cores from glaciers provided a wealth of new data on the history of temperature, atmospheric composition, ocean levels, and more.

Two cores, each over 3,000 meters deep, were drilled in Greenland early in the 1990s—they provided a record of conditions going back 100,000 years. Later in the decade, a French-Russian team working in the Vostok region of Antarctica extracted and analyzed a core that was 420,000 years old at its deepest point. Data from the Vostok study, published in 1999, has been described as "arguably among the most important produced by the global change scientific community in the twentieth century."[5] Subsequent drilling has extended the record to 800,000 years. It is no exaggeration to say that this research has revolutionized our understanding of Earth's past—and consequently, that it has revolutionized our understanding of Earth's present and future.

It has been known since the 1850s that small amounts of carbon dioxide in the atmosphere help to control Earth's temperature—CO_2 lets sunlight in, but won't let heat out. If the greenhouse effect did not exist, Earth's average temperature would likely be about 35°C colder than it is now, far colder than in the most extreme ice ages. We now also know that carbon dioxide constantly cycles between atmosphere and oceans, keeping the overall levels roughly stable.

It has also been long known that the angle at which sunlight hits Earth changes slightly over periods of approximately 100,000, 40,000, and 20,000 years—cycles produced by complex combinations of very slow changes in the shape of Earth's orbit and the tilt and orientation of the planet's axis. Climatologists have long believed that these Milankovitch cycles (named after the Serbian engineer who painstakingly calculated them in the 1920s) must play a role in the coming and going of ice ages, but the solar energy changes involved are simply too small to have had so much effect by themselves.

Detailed analysis of the composition of 800,000 years of Antarctic ice has now shown that the two apparently separate processes— wobbles in space and the terrestrial carbon cycle—are in fact closely linked as fundamental components of the Earth System. To oversimplify: the small amounts of cooling or warming caused by Milankovitch cycles act as triggers that cause CO_2 to be absorbed or released by the oceans, producing "changes that are abrupt and out of all proportion to the changes in incoming solar radiation."[6]

At the Amsterdam Global Change conference in 2001, the chair of the IGBP's Scientific Committee, Berrien Moore, pointed out that the cycles found in the Vostok ice core show a remarkably consistent pattern over hundreds of thousands of years:

> The repeated pattern of a 100 ppmv [parts per million by volume] decline in atmospheric CO_2 from an interglacial value of 280 to 300 ppmv to a 180 ppmv floor and then the rapid recovery as the planet exits glaciation suggests a tightly governed control system with firm stops at 280–300 and 180 ppmv. There is a similar CH_4 [methane] cycle between 320– 350 ppbv [parts per billion by volume] and 650–770 ppbv in step with temperature.[7]

IGBP executive director Will Steffen wrote that "no record is more intriguing than the rhythmic 'breathing' of the planet as revealed in the Vostok ice core records." The "remarkably regular planetary metabolic pattern embodied in the Vostok ice core" provided "a fascinating window on the metabolism of Earth over hundreds of thousands of years."[8]

The exact mechanisms of this "tightly governed control system" are still not fully understood, but there is no doubt that atmospheric CO_2 is the control knob on Earth's thermostat.

External factors can disrupt these cycles. About 56 million years ago, for example, a massive release of buried carbon dioxide, probably triggered by super-volcanoes or a comet collision, overwhelmed the normal process, increasing global temperatures by 5 to 9°C in a geological instant. It then took about 200,000 years for the excess CO_2 to be reabsorbed, and for temperatures to return to normal.[9]

The amount of CO_2 released in that episode was about equal to what will be produced if we burn all remaining reserves of coal, oil, and natural gas. Today's situation is different in many ways, so we should not expect a replay, but one important similarity should be noted. As Berrien Moore went on to say, atmospheric CO_2 levels are now far out of their normal range:

> Today's atmosphere, imprinted with the fossil fuel CO_2 signal, stands at nearly 100 ppmv above the previous "hard stop" of 280–300 ppmv. The current CH_4 value is even further (percentage-wise) from its previous interglacial high values. In essence, carbon has been moved from a relatively immobile pool (in fossil fuel reserves) in the slow carbon cycle to the relatively mobile pool (the atmosphere) in the fast carbon cycle.[10]

Figure 4.1 (page 65) illustrates the point. As the IGBP says in *Global Change and the Earth System*, "Human-driven changes are pushing the Earth System well outside of its normal operating range." And as climate change historian Spencer Weart says, learning the causes of ice ages showed that "the system is delicately poised, so that a little stimulus might drive a great change."[11] Burning fossil fuels has disrupted the carbon cycle, and global warming is an inevitable result—the questions are: how much, and how fast?

Tipping Points

The transformation of quantity into quality has been a fundamental postulate of dialectics for two centuries. Hegel stated and explained it as a law of thought; Marx and Engels applied it to the material world. Small changes accumulate, creating ever greater complexity, until the object or being or system suddenly shifts from one state to a radically different one, in what is often called a *phase change*. Water is a liquid until its temperature reaches 100°C when it becomes a gas. Overfishing produces large catches, right up until the fish population abruptly collapses. Social and economic stresses accumulate gradually until a revolutionary upsurge imposes a new social order, qualitatively different from the old society.

Few scientists today are familiar with dialectics, and even fewer use it consciously, but the fundamental dialectical concept of the transformation of quantity into quality has been absorbed into scientific thought under labels such as emergence, quantum leaps, and punctuated equilibrium.

Colloquially, those transitions are called "tipping points," a term originally used by physicists for the point at which adding weight or pressure to a balanced object suddenly causes it to topple into a new position. In the Earth System, tipping points are not unusual—they are the norm.

> Until a few decades ago it was generally thought that large-scale global and regional climate changes occurred gradually over a timescale of many centuries or millennia, scarcely perceptible during a human lifetime. The tendency of climate to change relatively suddenly has been one of the most surprising outcomes of the study of earth history.[12]

Despite this change in the scientific understanding of the climate, in most accounts, including the reports of the Intergovernmental Panel on Climate Change, there is an unspoken assumption that climate change will be gradual. The twenty-first century will be a warmer, stormier, and less biodiverse version of the twentieth—less pleasant, but not fundamentally different. As research commissioned by the U.S. National Research Council points out,

FIGURE 4.1: Global Climate Change

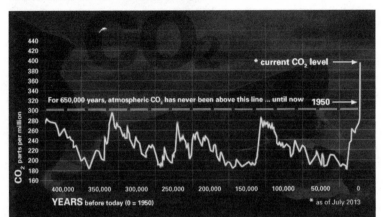

450,000 years of atmospheric CO_2. The dotted line indicates the upper bound of natural CO_2 variation as found in the Vostok Ice Core. By 1945, CO_2 was 25 parts per million above the preindustrial level; in 2015 it was 120 ppm above. Source: NASA, http://climate.nasa.gov/climate_resources/24/.

that assumption leads to particular conclusions about society's ability to respond to change:

> Many projections of future climatic conditions have predicted steadily changing conditions, giving the impression that communities have time to gradually adapt, for example, by adopting new agricultural practices to maintain productivity in hotter and drier conditions, or by organizing the relocation of coastal communities as sea level rises.

But the authors emphasize that the actual experience could be very different:

> The scientific community has been paying increasing attention to the possibility that at least some changes will be abrupt, perhaps crossing a threshold or "tipping point" to change so quickly that there will be little time to react. This concern is

reasonable because such abrupt changes—which can occur over periods as short as decades, or even years—have been a natural part of the climate system throughout Earth's history. The paleoclimate record—information on past climate gathered from sources such as fossils, sediment cores, and ice cores—contains ample evidence of abrupt changes in Earth's ancient past, including sudden changes in ocean and air circulation, or abrupt extreme extinction events.[13]

Many Earth System scientists argue that abrupt environmental change is not only possible, but virtually certain:

> In reality, Earth's environment shows significant variability on virtually all time and space scales. . . . Nonlinear, abrupt changes in key environmental parameters appear to be the norm, not the exception, in the functioning of the Earth System. Thus, global change is not likely to be played out as a steady or pseudo-linear process under any conceivable scenario but will almost surely be characterized by abrupt changes for which prediction and adaptation are very difficult.[14]

The Imbalance of Nature

The idea that the natural world is fundamentally stable and unchanging has a long history. In its oldest version, it is religious: God created a perfect world, and if humans disturbed that perfection, God would in time restore it. A secular equivalent was expressed in 1864 by the pioneering U.S. naturalist George Perkins Marsh:

> In countries untrodden by man the proportions and relative positions of land and water, the atmospheric precipitation and evaporation, the thermometric mean, and the distribution of vegetable and animal life are subject to change only from geological influences so slow in their operation that the geographical conditions may be regarded as constant and immutable.[15]

That view remains influential: one of the most quoted passages in all naturalist literature is Aldo Leopold's 1949 call for a "land ethic" based on preserving the "integrity, stability, and beauty of the biotic community."[16] It never occurred to Leopold, nor has it occurred to most of his contemporary admirers, that the natural world might be *inherently unstable*, subject to rapid change even in the absence of human activity.

Of course, naturalists had been aware since the mid-nineteenth century that glaciers on Earth had at least once advanced to cover much of the world with ice, and that animals now unknown had once walked the Earth, but changes of that magnitude were believed to occur extremely slowly, and to be of no relevance to human history and activity. Like the painted backdrop in a stage play, the natural world was the unchanging context, not an active player in any human drama.

That view is no longer tenable. Scientific research now shows that even in times of relative stability, like the Holocene, the Earth System is constantly changing on every scale of space and time, and that the most drastic changes often occur with remarkable speed.

Climate Chaos

In *Global Change and the Earth System*, Will Steffen and his colleagues wrote:

> The behavior of the Earth System is typified not by stable equilibria but by strong non-linearities, whereby relatively small changes in a forcing function can push the System across a threshold and lead to abrupt changes in key functions. Some of the modes of variability noted above contain the potential for very sharp, sudden changes that are unexpected given the relatively small forcing that triggers such changes. . . . The potential for abrupt change is a characteristic that is extremely important for understanding the nature of the Earth System. The

existence of such changes has been convincingly demonstrated
by paleo-evidence accumulated during the past decade.[17]

Figure 4.2, adapted from a study of ice-core data by scientists
at the Potsdam Institute for Climate Impact Research,[18] shows the
average annual temperature in Greenland over the past 100,000
years. Our current epoch, the Holocene, is the nearly flat segment
at the top right.

Ninety percent of the time shown in that graph was the end of
the Pleistocene, a 2.6 million-year-long epoch characterized by
repeated glaciations and interglacial retreats: the global climate
was not only cold, it was extremely variable. Modern humans
walked the Earth for the entire time shown in this graph, but until
the Holocene they all lived in small nomadic groups of hunter-
gatherers. Climate historian William J. Burroughs, who calls that
time the "reign of chaos," argues compellingly that so long as rapid
and chaotic climate change was the norm, agriculture and settled
life were impossible, even in parts of the world that the glaciers
never reached. To succeed, agriculture needs not just warm seasons, but a *stable and predictable climate*—and indeed, in just a
few thousand years after the Holocene began, humans on five continents independently took up farming as their permanent way of
life. "Once the climate had settled down into a form that is in many
ways recognizable today, all the trappings of our subsequent development (agriculture, cities, trade, etc.) were able to flourish."[19]

For 11,700 years, the average annual global temperature has not
varied up or down by more than one degree Celsius. But averages can conceal large variations: despite being warm and stable
on average, the Holocene has not been an unmitigated climate
paradise. That one degree average variation included uncounted
droughts, famines, heat waves, cold snaps and intense storms—
extreme weather events in which millions of people died.

The Pleistocene was far worse: temperature variations were five
to ten times greater than anything humanity has experienced since.

FIGURE 4.2: Average Annual Temperature in Greenland over the Past 100,000 Years.

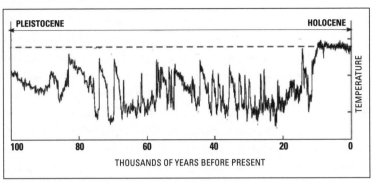

Furthermore, as geologists Jan Zalasiewicz and Mark Williams describe, the transition from Pleistocene cold to Holocene warmth was itself an abrupt and chaotic process;

> Coming from the Glacial Maximum, temperatures suddenly rose, 14,700 years ago, with the average temperature of the North Atlantic and surrounding areas increasing rapidly by some 5°C (over Greenland, the temperature hike approached 20°C). Temperatures remained around these levels for nearly two millennia—and then fell suddenly by a similar amount, as the whole region went into a deep freeze once more.

The new cold period, known to geologists as the Younger Dryas, lasted a thousand years. "Then, 11,700 years ago, temperatures suddenly soared again in another climate transformation—only this time the warm temperatures stayed, and this transition marks the beginning of the Holocene."[20]

How sudden is sudden? Each of the temperature jumps took a few decades, almost no time in geological terms, and not much time in human terms. More astonishing, in both cases the change in atmospheric circulation that drove the warming "seems to have been accomplished in something between *one and three years*."[21]

Holocene to Anthropocene

In 1999, the first scientists to study the Vostok ice core reported with surprise that "the Holocene, which has already lasted 11,000 years, is, by far, the longest stable warm period recorded in Antarctica during the past 420,000 years."[22] Will the transition to Anthropocene conditions be gradual, or can we expect sudden shifts comparable to those that initiated the Holocene?

The specific causes of past climate chaos are unlikely to repeat, so the pattern of change will certainly be different. But the most significant difference between then and now is the unprecedented impact of human activity in the past sixty years—and that makes it very likely, as a team headed by the noted U.S. biologist Anthony D. Barnosky concluded, that Earth is "approaching a state shift."

> Comparison of the present extent of planetary change with that characterizing past global-scale state shifts, and the enormous global forcings we continue to exert, suggests that another global-scale state shift is highly plausible within decades to centuries, if it has not already been initiated.[23]

If that occurs, the relative stability of the Holocene could be replaced by a new and unprecedented climate state, unlike anything any human society has experienced. And, as geoscientist Richard B. Alley points out, the transition is likely to be fast:

> Large, abrupt climate changes have repeatedly affected much or all of the earth, locally reaching as much as 10°C change in 10 years. Available evidence suggests that abrupt climate changes are not only possible but likely in the future, potentially with large impacts on ecosystems and societies. . . .
>
> Surprising new findings [show] that abrupt climate change can occur when gradual causes push the earth system across a threshold. Just as the slowly increasing pressure of a finger eventually flips a switch and turns on a light, the slow effects

of drifting continents or wobbling orbits or changing atmospheric composition may "switch" the climate to a new state. And, just as a moving hand is more likely than a stationary one to encounter and flip a switch, faster earth-system changes—whether natural or human-caused—are likely to increase the probability of encountering a threshold that triggers a still faster climate shift.[24]

Alley says that abrupt changes are most likely to occur in times when the climate system is under extreme stress, as it is today:

Abrupt climate changes were especially common when the climate system was being forced to change most rapidly. Thus, greenhouse warming and other human alterations of the Earth System may increase the possibility of large, abrupt, and unwelcome regional or global climatic events. The abrupt changes of the past are not fully explained yet, and climate models typically underestimate the size, speed, and extent of those changes. Hence, future abrupt changes cannot be predicted with confidence, and climate surprises are to be expected.[25]

Planetary Boundaries

In the first six years of the twenty-first century, two important new concepts emerged from Earth System science: the *Anthropocene*, and the *Great Acceleration*. In 2007, another investigation began "to identify which of Earth's processes are most important to maintaining the stability of the planet as we know it," and determine what must be done "to maintain Holocene-like conditions on Earth, now that we, in the Anthropocene, have become a global force of change."[26] That project led to a third key concept: *Planetary Boundaries*.

The project was initiated by the Stockholm Resilience Center, and headed by environmental scientist Johan Rockström. The list of participants reads like a who's who of Earth System science:

Rockström himself, Will Steffen, Paul Crutzen, James Hansen, Katherine Richardson, and some two-dozen more, from Sweden, the United Kingdom, the United States, and Australia. They wrote:

> We looked at one process after another, exploring interactions between them, and seeking to characterize the conditions needed to remain in a stable Holocene-like state. For each system, we made the best possible attempt, based on the latest scientific evidence at hand, to quantify the biophysical limits outside which the system might flip into a different, and for us undesired, state.[27]

The first results of the project were published in the journal *Ecology and Society* in September 2009, in a paper titled "Planetary Boundaries: Exploring the Safe Operating Space for Humanity." A less technical summary was published simultaneously in *Nature*.[28]

The authors began with the fact that all of humanity's history since we invented agriculture has taken place in the relatively warm and climatically stable period since glaciers last retreated from most of Earth. The goal was to quantify "the safe limits outside of which the Earth system cannot continue to function in a stable, Holocene-like state." As they wrote in a subsequent update:

> The human enterprise has grown so dramatically since the mid-20th century that the relatively stable, 11,700-year-long Holocene epoch, the only state of the planet that we know for certain can support contemporary human societies, is now being destabilized. In fact, a new geological epoch, the Anthropocene, has been proposed.
>
> The precautionary principle suggests that human societies would be unwise to drive the Earth System substantially away from a Holocene-like condition. A continuing trajectory away from the Holocene could lead, with an uncomfortably high probability, to a very different state of the Earth System, one

that is likely to be much less hospitable to the development of human societies.[29]

The 2009 paper identified nine Planetary Boundaries, ecological processes that maintain the "safe operating space for humanity" that has characterized the entire history of civilization until now. Disruption of any of these processes could lead to "linear, abrupt environmental change within continental- to planetary-scale systems." The following is a brief summary.

New names and parameters were introduced for some of the processes in 2015. For reference, the previous names of planetary boundaries are given in parentheses. For more detail, see Table 4.1 (page 75), and the Stockholm Resilience Center website *http://www.stockholmresilience.org*.

Climate change. The volume of greenhouse gas in the atmosphere is the highest it has been for hundreds of thousands of years: the planetary average passed 400 parts per million in 2015.

Changes in biospheric integrity (was: *rate of biodiversity loss*). It has been estimated that species are now going extinct at a rate about 1,000 times greater than in preindustrial times.

Biogeochemical flows (was: *nitrogen and phosphorus*). Fertilizers containing nitrogen and phosphorus, both essential for plant growth, are widely used in modern agriculture. As much as 50 percent of nitrogen ends up in lakes, rivers, and oceans, where it can cause abrupt ecosystem changes such as the notorious "dead zone" in the Gulf of Mexico. In future this category may include other elements.

Stratospheric ozone depletion. In the 1970s, scientists learned that widely used chemicals were destroying the ozone that blocks harmful ultraviolet radiation from reaching the surface of Earth. (See chapter 6 for further discussion.)

Ocean acidification. A proportion of CO_2 emissions dissolves in seawater, making it much more acidic than in preindustrial times. This can interfere with the growth and survival of corals, many shellfish, and plankton, causing the collapse of essential food webs and drastic reductions in fish and marine mammal populations.

Freshwater use. Heavy withdrawals for agricultural and industrial uses are depleting major aquifers, while melting glaciers are eliminating the source water of many rivers. Current global water use by humans totals about 2,600 cubic kilometers a year, which is below the global limit, but withdrawals are above regional limits in many areas.

*Land-system change (*was: *land-use change).* About 42 percent of all ice-free land is currently used for farming: that land formerly supported 70 percent of the world's grasslands, 50 percent of savannahs, and 45 percent of temperate deciduous forest. Loss of this land reduces biodiversity and has negative effects on Earth's climate and water systems.

Atmospheric aerosol loading. Most of what is usually called "air pollution" consists of microscopic particles and droplets called aerosols. Inhaling them causes about 7.2 million deaths per year. In addition, they have a direct effect on climate, most notably by significantly reducing monsoon activity.

*Introduction of novel entities (*was: *chemical pollution).* There are over 100,000 chemicals, nanomaterials, and plastic polymers in commercial use today. For almost all, very little is known about their individual or combined effects on human or ecosystem health. The name was changed to allow inclusion of genetically modified organisms and radioactive materials.

Planetary boundaries are not the same as tipping points. They can be compared to guardrails on mountain roads, which are positioned to prevent drivers from reaching the edge, not on the edge itself.

> A zone of uncertainty, sometimes large, is associated with each of the boundaries. . . . This zone encapsulates both gaps and weaknesses in the scientific knowledge base and intrinsic uncertainties in the functioning of the Earth System. At the "safe" end of the zone of uncertainty, current scientific

TABLE 4.1: Planetary Boundaries

Earth System Process	Control Variables	Planetary Boundary (zone of uncertainty)	Current Value of Control Variables
Climate Change	Atmospheric CO_2 concentration, parts per million.	350 (350-450)	396.5
	Energy imbalance at top of atmosphere. Watts per square meter.	+ 1.0 (+1.0–1.5)	2.3 (1.1–3.3)
Biosphere Integrity	Genetic Diversity: Number of extinctions per year per million species	10 (10–100) Aspirational goal: 1	100–1000
	Functional Diversity Biodiversity Intactness Index (interim variable)	90% (90%-30%)	84.4% (Southern Africa only)
Novel Entities	To be determined		
Stratospheric Ozone Depletion	Stratospheric Ozone concentration. Dobson Units	5% below preindustrial level of 290. (5%–10%)	Approx. 200 in Antarctica, spring only.
Ocean Acidification	Aragonate surface saturation	80% or less of preindustrial (80%–70%)	Approx. 85%
Biogeochemical Flows (Phosphorus and Nitrogen Cycles)	Global P cycle: flow from freshwater systems into the ocean. Million tonnes/year	11 (11–100)	22
	Regional P cycle: flow from fertilizers to erodible soils. Million tonnes/year	3.72 (3.72–4.84)	Approx 14
	Global N cycle: industrial and intentional biological fixation of N. Million tonnes/year	44 (44–62)	Approx 150
Land-system Change	Global: area of forested land as % of original forest cover	75% (75% –54%)	62%
	Biome: area of forested land as % of penitential forest	Tropical: 85%(85%–60%) Temperate: 50%.(50%–30%) Boreal: 85% (85%–60%)	
Freshwater Use	Global: Maximum blue water consumption. Cubic kilometers/year	4,000 (4,000–6,000)	Approx. 2,600
	Basin: Blue water withdrawal as % of mean monthly runoff	Low-flow months: 25% (25%–55%) Medium-flow months: 30% (30%–60%) High-flow months: 55% (55%– 85%)	
Atmospheric Aerosol Loading	Global: Aerosol Optical Depth (AOD)		
	Regional: AOD as seasonal average. (Data for south Asian monsoon only.)	0.25 (0.25–0.50) Warming AOD less than 10% of total	0.33

Adapted from Steffen et al., "Planetary Boundaries: Guiding Human Development"; and Rockström and Klum, *Big World, Small Planet.* Updated 2015.

knowledge suggests that there is very low probability of cross-ing a critical threshold or significantly eroding the resilience of the Earth System. Beyond the "danger" end of the zone of uncertainty, current knowledge suggests a much higher prob-ability of a change to the functioning of the Earth System that could potentially be devastating for human societies.[30]

The 2009 article tentatively assigned numeric limits for seven of the planetary boundaries, and showed that three of those—climate change, nitrogen pollution, and biodiversity loss—were already in the danger zone, and three more were heading that way. An update published in 2015 redefined the definitions and limits in the light of continuing research: it concluded that four out of nine boundaries have now been passed. "Two are in the high risk zone (biosphere integrity and interference with the nitrogen and phos-phorus cycles), while the other two are in the danger zone (climate change and land-use change)."[31]

It is important to bear in mind that although the planetary boundaries are defined separately, in reality they are tightly linked.

We do not have the luxury of concentrating our efforts on any one of them in isolation from the others. If one boundary is transgressed, then other boundaries are also under serious risk. For instance, significant land-use changes in the Amazon could influence water resources as far away as Tibet. The cli-mate-change boundary depends on staying on the safe side of the freshwater, land, aerosol, nitrogen-phosphorus, ocean, and stratospheric boundaries. Transgressing the nitrogen-phosphorus boundary can erode the resilience of some marine ecosystems, potentially reducing their capacity to absorb CO_2 and thus affecting the climate boundary.[32]

Scientists associated with the IGBP have warned that abrupt cli-mate change is particularly dangerous: "Societies can have little or no warning that a forcing factor is approaching such a threshold, and by the time that the change in Earth System functioning is observed, it will likely be too late to avert the major change."[33]

Wallace Broeker, who's often called the grandfather of climate change, expresses it more dramatically: "The paleoclimate record shouts out to us that, far from being self-stabilizing, the Earth's climate system is an ornery beast which overreacts even to small nudges."[34]

Atmospheric carbon dioxide is now over 400 parts per million, compared to a maximum of 280 during the wildest climate swings of the Pleistocene. That is much more than a small nudge, so no one should be surprised if the ornery beast strikes back violently, again and again. Nor should anyone be surprised if the result is a world unlike anything humanity has ever seen.

5

The First Near-Catastrophe

Activities damaging to the environment may be relatively harmless when introduced on a small scale; but when they come into general use and spread from their points of origin to permeate economies on a global scale, the problem is radically transformed. This is precisely what has happened in case after case, especially in the half-century following the Second World War, and the cumulative result is what has become generally perceived as the environmental crisis.

—PAUL M. SWEEZY[1]

Sudden environmental transitions have been normal and frequent for millions of years. Now, with human activity adding stresses to the normal processes of global change, the possibility of abrupt shifts with potentially catastrophic results is greater than ever. That is not just speculation about what *might* happen. Late in the twentieth century, the Earth System crossed a tipping point and was headed toward disaster. The completely unexpected threat was identified only when the crisis was already far advanced, and even then decisive action was delayed. The impact is still being felt, and will continue to cause pain and premature death for thousands, for many decades to come.

The crisis, popularly known as the hole in the ozone layer, was widely discussed in the 1980s and 1990s, but lately it has all but vanished from environmental discussions. Recent books on the global environmental crisis either omit it entirely or mention it only briefly. That's unfortunate, because the story of the ozone layer teaches important lessons. In particular, it illustrates what Earth System scientists mean when they say that in the Anthropocene human activity is *overwhelming the great forces of nature* with potentially catastrophic results. The ozone crisis was the first major demonstration of that, the first near-catastrophe of the Anthropocene.

———————✿———————

Ozone (O_3) is a form of oxygen in which each molecule contains three oxygen atoms instead of the usual two. It's rare—there are roughly 3 ozone molecules in the atmosphere for every 10 million oxygen molecules—and almost all of it is in the upper atmosphere, between 15 and 30 kilometers above sea level.

In the late 1800s, scientists found that sunlight reaching Earth contained far less ultraviolet radiation than expected: all of the shortest wavelengths and most of the middle-range wavelengths emitted by the Sun are blocked by the ozone in the upper atmosphere. In 1931, a British chemist, Sydney Chapman, showed that this was not a passive process, but a complex cycle in which interaction with ultraviolet radiation converts oxygen into ozone and ozone into oxygen. The cycle ensures that the proportions of oxygen and ozone remain roughly constant, and that most ultraviolet energy is absorbed long before it reaches Earth's surface.

This might be just a matter of academic interest, were it not for one critical fact: if that ultraviolet light were not blocked, it would be catastrophic for life on Earth.

> UV-C, the shortest-wavelength light . . . is sufficiently lethal to use in sterilizations. . . . UV-B radiation . . . kills phytoplankton, the basis of oceanic food chains. It affects all photosynthesis in green plants. In humans it causes cataracts and other eye

ailments, suppresses immune system response, and in suscep-
tible people causes skin cancer.[2]

The thin layer of ozone, far above our heads, is a shield against
UV-B and UV-C radiation that would be damaging in the short
term and ultimately deadly.

Making Refrigeration Safe

In one of those coincidences that you would not believe in a movie,
just when Chapman was discovering how the ozone layer works, a
U.S. corporation was introducing the product that would eventually threaten to destroy it. At the 1930 meeting of the American
Chemical Association, Thomas Midgley Jr. introduced a product
that would make home refrigerators practical and safe—and eventually bring Earth to the brink of disaster.

During the 1920s, the number of U.S. homes with electricity
grew from 25 to 80 percent. Giant manufacturers like General
Motors, General Electric, and Westinghouse saw an opportunity
to create a mass market for electrical appliances, and they were
urged on by electrical utilities because more devices meant higher
electricity bills. Electric irons, stoves, and washing machines were
among the first products to be promoted as part of what historian
Ruth Schwartz Cowan justly calls "the industrial revolution in the
home."[3] Many manufacturers saw an opportunity to replace home
iceboxes with electric refrigerators, but adoption was slow, in part
because the first machines were expensive, and in part because
they were downright dangerous, using toxic and inflammable
gases such as ammonia, sulfur dioxide, or methyl chloride as
refrigerants. Even if they didn't kill, refrigerator gas leaks smelled
bad and spoiled food.

This was the era when invention by individual tinkerers was
being supplanted by research laboratories that invented on com-
mand. In large industry, as Marx wrote when the process had
barely begun, "conscious and planned applications of natural

science, divided up systematically in accordance with the particular useful effect aimed at in each case," became the norm.[4] One of the most successful of such operations was Dayton Engineering Laboratories Company (Delco), organized by Charles Kettering in 1909 and acquired by General Motors Corporation in 1919.

In 1928, General Motors asked Delco to develop a safe refrigerant for GM's Frigidaire line of home refrigerators. Kettering assigned the project to Thomas Midgley, who had previously invented the first anti-knock lead additive for gasoline and was now trying to develop synthetic rubber for tires. According to corporate legend, his team came up with a new refrigerant in just three days—a family of highly stable synthetic gases called chlorofluorocarbons, or CFCs. Midgley liked to demonstrate that CFCs were non-toxic and non-inflammable by inhaling a lungful and then blowing out a candle.

Production of the new gas was assigned to GM's principal owner, the chemical giant DuPont, which took less than two years to move the invention into industrial production. In 1930 it started shipping CFCs from a new purpose-built factory in New Jersey, using the brand name Freon.

Freon immediately gave GM's Frigidaire refrigerators such a competitive edge that by the mid-1930s every major refrigerator maker was either buying Freon from DuPont or manufacturing it under license. Sales were strong despite the Depression, but really took off after the Second World War, when hundreds of millions of CFC-based refrigerators and air-conditioning units were sold around the world. CFC sales went beyond refrigeration: because it was odorless, tasteless, and non-toxic, and could be compressed easily, it could be used as a propellant for almost any product that could be sprayed from a can. By the 1970s, aerosol spray products—insecticides, hair spray, deodorants, engine lubricants, perfume, paint, and much more—accounted for more than half of all CFCs used in the United States.[5] This success meant that ever more CFCs were released into the atmosphere—about 20,000 tons in 1950, soaring to about 750,000 tons in 1970.[6]

Until 1974, CFCs were not on the environmental agenda because they appeared to be the closest thing to inert and harmless. They were non-toxic and non-inflammable precisely because they did not interact with other substances or break down into other chemicals. That situation changed as a result of three unrelated studies by scientists who had no intention of changing Freon's unsullied reputation as a useful product that did no harm.

CFCs and Ozone

In 1970–71, Dutch atmospheric chemist Paul Crutzen showed that the chemistry of the ozone layer was more complex than had been believed. In particular, he showed that nitrogen oxides played an important role in keeping ozone levels constant. He warned that any increase in the nitrogen oxides in the upper atmosphere—for example, by the then-proposed fleet of high-flying supersonic jetliners—could deplete the ozone layer and expose Earth to increased ultraviolet radiation. His analysis laid the basis for all subsequent scientific study of the ozone layer.

In January 1971, British scientist James Lovelock was looking for evidence that industrial smog could travel long distances. He identified CFCs, not as a source of smog, but as a marker that would show how human-made products move in the atmosphere. Using a device he invented for measuring trace gases, he took frequent air samples while on a scientific expedition from Britain to Antarctica. In January 1973, in the journal *Nature*, he reported that he had found CFCs in every single sample, "wherever and whenever they were sought." The explanation was simple: precisely because they don't easily break down or combine with anything else, practically every CFC molecule ever made is still in the air, somewhere. It was Lovelock's view, however, that "the presence of these compounds constitutes no conceivable hazard."[7]

A few months later, a chemistry professor at the University of California, Sherwood Rowland, suggested to Mario Molina, his newly hired postdoctoral research assistant, that it might be interesting to determine where all those CFC molecules were going. Neither had a background in atmospheric chemistry, so this was an opportunity to spread out, so to speak. They found that there were no natural "sinks" for CFCs in the lower atmosphere—they were not washed out by rain, and they did not combine with any other substances. But they found any CFC molecules that reached the stratosphere—which almost all would, eventually—were destroyed by high-energy ultraviolet radiation, in a process that released chlorine into the ozone layer. And what happened to the chlorine? To their shock, Rowland and Molina found that it would act as a high-powered catalyst, producing a chain reaction in which every chlorine atom could destroy up to 100,000 ozone molecules. The ozone layer was under attack.

In 1974, Rowland and Molina published their findings in *Nature* and reported them to the American Chemical Society. They estimated that 1 percent of the ozone layer had already been destroyed, and that if CFC production continued, 5 to 7 percent would be gone by 1995, and 50 percent by 2050. Even a 10 percent reduction could increase cases of skin cancer in the United States by 80,000 a year. To protect life on Earth, they argued, chlorofluorocarbons should be banned.[8]

The chemical industry, led by DuPont, went into all-out "deny and delay" mode. They insisted, correctly, that Rowland and Molina had no evidence that any ozone was actually being destroyed. *This is just theory, it might be wrong, regulation will destroy jobs, more study is needed, there is no need for urgent action*—all the arguments we have heard since in support of tobacco and fossil fuels were marshalled to protect CFC profits. Although most CFC-based aerosol products were banned in the United States, Canada, and Scandinavia by the late 1970s, they remained on the market elsewhere, and production for other uses continued to grow. Within eight years total CFC production was higher than before the aerosol ban.

A few scientists, including Rowland and Molina, continued to campaign for strong action, but the dominant trend among

scientists who addressed the CFC issue in the 1980s was conservative. Most notably, in 1984 the U.S. National Academy of Sciences,
which had previously predicted that ozone would decline 16.5
percent by 2100, reduced that forecast to 2–4 percent. Such conservatism was well received by the anti-environment ideologues
in the Reagan administration, and gave credibility to the chemical
industry's wait-and-see policy. DuPont stopped research on alternatives to CFCs, and the U.S. Environmental Protection Agency
never introduced a promised second phase of CFC regulations.

A key factor in the CFC industry's ability to avoid a ban was the
absence of hard scientific data showing a decline in ozone levels.
Satellites began measuring atmospheric ozone in the late 1970s,
but without historical data it was impossible to know whether any
changes were significant or just natural fluctuations. As DuPont
officials said in 1979, "No ozone depletion has ever been detected
despite the most sophisticated analysis. . . . All ozone-depletion
figures to date are computer projections based on a series of uncer-
tain assumptions."[9] The theory was strong, but untested, and any
scientists who predicted rapid ozone depletion in the absence of
evidence were vulnerable to charges of crying wolf.

Until 1985, no one involved in the CFC debates appears to have
been aware that a detailed historical record of ozone levels actually existed. Since 1957, the British Antarctic Survey, working on
a shoestring budget of about $18,000 a year, had been measuring
the ozone layer from the Halley Bay Observatory in Antarctica.
In 1982, the longtime head of the project, Joe Farman, noticed an
unusual drop in ozone levels, but thought it might be an equipment malfunction. When new equipment and observations at
another location 1,000 miles away showed the same thing, he
decided to tell the world. Farman and three associates published
their findings in *Nature*, in May 1985.

Sherwood Rowland later described Farman's article as "the
greatest surprise in the CFC-ozone story."[10] Beginning in 1979,
springtime (October in the Southern Hemisphere) levels of ozone
above Antarctica had gone into steep decline. The October level

fell from around 300 Dobson units in the 1970s to as low as 125 in the mid-1980s—nearly a 60 percent drop. No one had predicted a decline anywhere close to that speed and magnitude; a recent NASA report had predicted that ozone might decline 5 to 9 percent by 2050.

But if Antarctic levels were as low as Farman said, why hadn't satellites seen the same thing? Answer: they had, but the computers that analyzed the data had been programmed to treat unusually low measurements as errors, and ignored them. NASA's review of the raw data confirmed that the minimum springtime level of ozone over Antarctica had fallen 40 percent between 1979 and 1984. A hole in the ozone layer, twice as large as the Continental United States, had developed in just five years.

CFCs had been entering the atmosphere in ever-increasing amounts since the early 1930s. The pre-1980 measurements from Halley Bay undoubtedly included their effect on the ozone layer above Antarctica, but in 1979, what had been a gradual, linear process crossed a tipping point, becoming rapid and nonlinear. As dialectical logic would express it, quantitative change became qualitative change.

Over the next two years, intense scientific investigation confirmed that the ozone hole was real, and that it was caused by CFCs. The chemistry proved to be even more complex than previous studies had suggested: the extreme cold and high winds of the Antarctic winter radically accelerate the breakdown of CFCs and the destruction of ozone, and isolate the area from the rest of Earth's atmosphere. When winter ends, ozone from warmer latitudes flows in, so the amount of ozone in the global ozone layer is reduced.[11] By the mid-1990s, average stratospheric ozone levels in mid-northern and mid-southern latitudes had fallen 10 percent.

In September 1987, under heavy scientific and public pressure, twenty-seven countries and the European Economic Community signed the Montreal Protocol on Substances that Deplete the Ozone Layer, pledging to phase out most CFC production by 2000.

In 1988 DuPont abandoned its opposition and agreed to stop CFC production by century's end, and the rest of the chemical industry reluctantly fell into line.[12]

Ozone loss seems to have peaked in the Antarctic in 2006, and in the Arctic in 2011. The ozone layer will not recover until all CFCs are gone, a process that is likely to take most of the twenty-first century to complete. It has been estimated that by 2000, ozone depletion had caused well over a million cases of skin cancer and between ten and twenty thousand early deaths. Many thousands more will die before ultraviolet radiation returns to pre-CFC levels.[13]

The CFC-ozone crisis is a powerful example of what the new epoch means. Production of CFCs opened a massive hole in one of the globe's basic metabolic systems: it disrupted the complex inter-action between ultraviolet radiation and the ozone layer, a process that has protected life on Earth for billions of years, bringing it to the brink of collapse—and it did so with frightening speed. As a 2013 report to the U.S. government stated, "The Antarctic ozone hole represents an abrupt change to the Earth System. . . . It exemplifies the scope and magnitude of the types of impacts that abrupt changes from human activities can have on the planet."[14]

What's more, as Paul Crutzen said in his Nobel Prize lecture in 1995, "Things could have been much worse." If GM and DuPont had used bromine instead of chlorine, the resulting gas would have been just as effective as a refrigerant, but bromine is 50 to 100 times more effective at destroying ozone, and it does not require the deep Antarctic cold to operate effectively:

> If the chemical industry had developed organobromine compounds instead of the CFCs . . . then without any preparedness, we would have been faced with a catastrophic ozone hole everywhere and at all seasons during the 1970s, probably before the atmospheric chemists had developed the necessary knowledge to identify the problem and the appropriate techniques for the necessary critical measurements.[15]

Even with the chemistry as it was commercialized, it was far from inevitable that the disaster threatened by CFCs would be identified and stopped in time. If Joseph Farman had not continued measuring Antarctic ozone despite the absence of practical applications, if James Lovelock had not spent most of a year measuring CFC levels in order to prove a point about atmospheric circulation, if Sherwood Rowland had assigned a different project to his new research assistant—these and many other contingencies could have led to a very different outcome.

After reviewing the history of CFCs and the ozone layer, Paul Crutzen commented, "I can only conclude that mankind has been extremely lucky."[16]

That's true—but "luck" in this case depended on capitalist profit. DuPont supported an international ban only because CFC profits were in steep decline and more profitable alternatives were nearly ready. As historians James Maxwell and Forrest Briscoe have shown:

> DuPont's decision to support a CFC ban was based on the belief that it could obtain a significant competitive advantage through the sales of new chemical substitutes because of its proven research and development capabilities to develop chemicals, its (limited) progress already made in developing substitutes, and the potential for higher profits in selling new specialty chemicals. . . . The international regulatory regime had the potential to transform one of DuPont's mature and only marginally profitable businesses into a more lucrative one.[17]

If that had not been the case, we might still be fighting the results of a CFC disinformation campaign comparable to big oil–funded climate change denial.

The ozone crisis is often cited as proof that capitalism can solve global environmental problems: if international negotiations could save the ozone layer, then why not the climate? This ignores the fact that the ozone crisis could be solved by a handful of companies and industries implementing a technical fix. In contrast, eliminating fossil fuels and greenhouse gas emissions will require a decades-long transformation of the global economy.

After years of trying to convince giant chemical corporations to stop producing CFCs, Sherwood Rowland concluded: "As far as industry is concerned, they have great difficulty looking more than ten years down the road."[18] That judgment, based on harsh experience with capitalist reality, is both accurate and frightening. The CFC-ozone crisis was the first near-catastrophe of the Anthropocene, but unless major changes are made, there will be many more.

6

A New (and Deadly) Climate Regime

Actions taken during this century will determine whether the Anthropocene climate anomaly will be a relatively short-term and minor deviation from the Holocene climate, or an extreme deviation extending over many thousands of years.

— U.S. NATIONAL RESEARCH COUNCIL[1]

In Paris in 2015, climate negotiators agreed that "well below 2°C above preindustrial levels," is an appropriate limit for any increase in global average temperature, and that a 1.5°C limit would be preferable. [2] Noted climate scientist James Hansen described that promise as "a fraud really, a fake. . . . It's just bullshit," because the Paris Agreement includes no concrete measures for ensuring that those limits are not exceeded.[3]

In the months before the Paris conference, 158 countries submitted "Intended Nationally Determined Contributions"—basically, announcements of the emission reductions they planned to voluntarily implement. Expert analysis of those INDCs showed that if every promise is kept to the letter, there is a 90 percent chance that the temperature will rise more than 2°C by 2100, and a 33 percent chance it will rise more than 3°C.[4] If business as usual continues—if the INDCs, like all previous pledges, prove to be just hot air—the global average temperature could be 4°C above the preindustrial level by 2080.

That doesn't sound like much. When I awoke on a recent August morning, the temperature outside was 19°C, and by noon, it was 25°C. That's a 6-degree jump in five hours or so, a pretty common experience in summer where I live in Canada. So why would we worry about an increase of 2 or 4 degrees by 2100? Mention that at a party in my neighborhood, and people are sure to say that they would be very happy if our Canadian winters were 4 degrees warmer!

It may be counterintuitive, but 4 degrees is actually a huge jump. During the last ice age, when kilometers of ice covered areas as far south as present-day Chicago and Berlin, the average global temperature was only 5 degrees Celsius (9 degrees Fahrenheit) cooler than today.

It is important to remember that average global temperatures conceal substantial variations in time and place. For example, the atmosphere is consistently cooler over oceans, so a 4-degree average global increase could mean a 6-degree or more increase on land and a 16-degree increase in the Arctic. In the tropics, the increase would likely be less than 4 degrees, but that smaller shift would be from very hot to extremely hot.

There is more to global warming than average thermometer readings: temperature change can cause dramatic shifts in weather patterns, biodiversity, and much more. Unless radical changes are made, the Anthropocene will be marked not just by warmth but by a new climate regime, very different from the 11,700 years of Holocene stability. This is not just speculation: the transition is well under way.

Figure 6.1 explains the role of standard deviation in evaluating climate variability.

Loading the Climate Dice

In September 2012, the prestigious journal *Proceedings of the National Academy of Sciences* published a study that examined how global warming to date has affected global climate. The results were remarkable: authors James Hansen, Makiko Sato, and Reto Ruedy showed that the frequency of extreme temperatures, particularly in the summer, has "changed dramatically in the past three decades."[5]

As a baseline, Hansen, Sato, and Ruedy examined the period 1951–80, before rapid global warming began and while the climate

FIGURE 6.1: Standard Deviation and Sigma Events

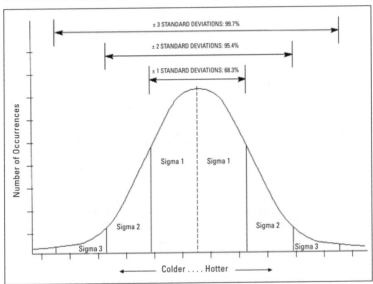

Scientists calculate the standard deviation to measure how much a set of observations is spread out from the average, or mean. If the data is normally distributed, about 68 percent of the observations will be within one standard deviation on either side the mean, and about 95 percent will be within two standard deviations. Anything beyond that will be possible, but rare. A graph of average July temperatures over many years might look like the Figure above.

The bell curve would be wider if the range of temperatures was greater, narrower if the range was smaller, and the width of each standard deviation would change similarly.

The mathematical symbol for standard deviation is the Greek letter sigma (σ). Observations that are more than two standard deviations from the mean are called Sigma-3 events. Much rarer are Sigma-4 events, Sigma-5 events, and so on. Temperatures in the Sigma-3 band will normally occur less than once a century, and those in the Sigma-5 band no more than once in several million years.

FIGURE 6.2: Hotter and More Extreme

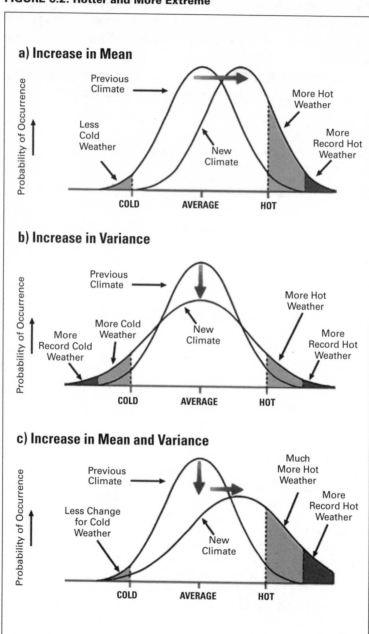

was still within the long-term Holocene range. They graphed average monthly temperatures for small (250 km × 250 km) areas covering the entire globe, and calculated the standard deviation for each measurement, to show how much year-to-year variation there was. As expected, the temperatures did not vary very much: almost every temperature point was within two standard deviations of the mean, Sigma-1 or Sigma-2. They then did the same for 1981–2010, and compared the results.

Since we know that Earth has been warmer, on average, since 1981, it is not surprising that the graphs have shifted to the right—the mean is warmer, and so are individual temperatures. The shocking discovery is that the graphs have also changed shape, because the amount of variation has increased. Hansen, Sato, and Ruedy write:

> The distribution of seasonal mean temperature anomalies has shifted toward higher temperatures and the range of anomalies has increased. An important change is the emergence of a category of summertime extremely hot outliers, more than three standard deviations (3σ) warmer than the climatology of the 1951–1980 base period. This hot extreme, which covered much less than 1% of Earth's surface during the base period, now typically covers about 10% of the land area.[6]

Figure 6.2 shows what happened.

In practice, this means that extreme heat waves—Sigma-3 events, which were virtually nonexistent in 1951–80, have become increasingly frequent. For example:

(Left) This schematic, adapted from the IPCC's 2001 report on the scientific basis for climate change, shows the effect on extreme temperatures when (a) the mean temperature increases, (b) the variance increases, and (c) when both the mean and variance increase. James Hansen's 2012 paper shows that (c) has in fact happened, producing weather that is simultaneously hotter and more prone to extremes.

- Europe, 2003: The hottest summer in at least 500 years killed over 70,000 people.
- Russia, 2010: The hottest summer since 1500 caused 500 wildfires near Moscow and cut grain harvests by 30 percent. Fifty-six thousand people died.
- United States, 2011: The most extreme July heat and drought since 1880 in Texas and Oklahoma. Damage was estimated at $13 billion.
- India, 2015: Twenty-five hundred people died in a heat wave in May, when temperatures went as high as 47.7°C (117°F).

Referring to events such as these, Hansen, Sato, and Ruedy write: "We can say with high confidence that such extreme anomalies would not have occurred in the absence of global warming."[7]

They compare the climate to colored dice. In the baseline decades, two sides were red for hot, two were blue for cold, and two were white for near-average—the odds of any month being red, blue, or white were equal. Now, the dice have been rigged: four sides are red, so hot comes up more often. But the analogy is breaking down, because we have to add a new category—*extremely hot.*

In a commentary, climatologists Thomas Karl and Richard Katz describe this study as a "simple and elegant" demonstration that "we are now more than 10-fold as likely to endure an extremely hot summer than we were in the decades 1951–1980."

> Hansen et al. have made the case that we are no longer waiting for evidence of global warming. It is clearly here now, affecting a wide variety of weather and climate events, and it will continue to grow as we burn more fossil fuels. . . . Even the apparently normal distribution of temperature can display non-normal behavior, and this can lead to extremes of even greater magnitude than might otherwise be expected.[8]

In short, an average temperature increase of less than 1 degree Celsius has already disrupted the global climate system, pushing

Earth's climate out of Holocene conditions—and this is only the beginning. As Hansen, Sato, and Ruedy warn, another 1 degree increase will have devastating consequences: "In that case, the further shifting of the anomaly distribution will make Sigma-3 anomalies the norm and Sigma-5 anomalies will be common."[9]

Other studies have come to similar conclusions. A 2012 paper examined the "exceptionally large number of record-breaking and destructive heat waves" in the first decade of this century. The authors found that "many lines of evidence—statistical analysis of observed data, climate modelling and physical reasoning—strongly indicate that some types of extreme event, most notably heat waves and precipitation extremes, will greatly increase in a warming climate and have already done so. . . . The evidence is strong that anthropogenic, unprecedented heat and rainfall extremes are here—and are causing intense human suffering."[10]

A study of extreme weather events between 1997 and 2012 concluded that "the available evidence suggests that the most 'extreme' extremes show the greatest change. This is particularly relevant for climate change impacts, as changes in the warmest temperature extremes over land are of the most relevance to human health, agriculture, ecosystems, and infrastructure."[11]

Others have found that global warming made the 2012–14 California drought significantly worse than it would have otherwise been, and that in Australia, where hot and cold records used to be set in about equal numbers, all-time hot weather records now outnumber cold records by 12 to 1.[12]

In short, the climate is not just getting warmer on average; *the climate pattern is skewing toward heat extremes.* That is critically important, because adapting to the new normal—if that is even possible—will require responding to extremes, not just averages. For human well-being and survival, the issue is not just how much the average ocean level rises, but how high the biggest storm surges are; not just what the average daily rainfall is, but how long the droughts last; not just how much warmer it gets on average, but how long and how deadly the heat waves become.

If one degree of global warming has made extreme weather significantly more probable, what can we expect in coming decades, as the first full century of the Anthropocene unfolds? A growing body of scientific research has focused on that question, and the answers are consistent: the extreme heat events of recent years will become more frequent and more severe as the century progresses.

In 2012, for example, a special IPCC report on the nature and likelihood of extreme events concluded that "it is virtually certain that increases in the frequency and magnitude of warm daily temperature extremes and decreases in cold extremes will occur through the twenty-first century at the global scale," and very likely that the length, frequency, and/or intensity of warm spells or heat waves will increase over most land areas. Under business-as-usual scenarios, "a 1-in-20-year hottest day is likely to become a 1-in-2-year event by the end of the twenty-first century in most regions."[13]

Mean Projections and Dangerous Change

The 2°C target adopted in Paris has been called a death sentence for billions of people in Africa, Asia, and Latin America. As British climatologists Kevin Anderson and Alice Bows-Larkin write, 2°C "represents a threshold, not between acceptable and dangerous climate change, but between dangerous and 'extremely dangerous' climate change."[14]

Nevertheless the 2°C pledge gives us a benchmark: if action on climate change depends on the voluntary actions of those governments, that is the *most* we can expect. The reality is likely to be somewhere between that and business as usual—that is, doing little or nothing to cut greenhouse gas emissions, a policy which will likely raise the global average temperature by 4°C or more over the preindustrial level before 2100.

The World Bank, which has invested billions of dollars in fossil fuel developments, is no altruistic friend of the environment. But it also invests in climate change adaptation and other climate-related projects, so perhaps it had those opportunities in mind when it commissioned the prestigious Potsdam Institute for Climate Impact Research (abbreviated PIK, from its German initials) to study and report on the impact of the business-as-usual option and the possibility of a 4-degree temperature increase. How much difference could two additional degrees make?

Whatever the World Bank's motives, it could not have expected the outcome. PIK itself has been a leader in climate research since 1992, and yet the Institute's director, Hans Schellnhuber, told the UN climate conference in Doha in 2013 that "even we, the researchers, were shocked by our findings."[15]

A New Normal

PIK's findings were published by the World Bank in three substantial reports, in 2012, 2013, and 2014, under the overall title *Turn Down the Heat*. The reports combined an extensive review of research by scientists around the world, with state-of-the-art computer modelling based on the IPCC's new emission scenarios, called "Representative Concentration Pathways" (RCPs).[16] They focused on two of these: RCP 2.6, under which immediate emissions cuts and "negative emissions" after 2050 may keep temperature increases below 2°C, and RCP 8.5, which is essentially a business-as-usual projection under which emissions and temperatures continue growing, which PIK projects will lead to global warming of about 4°C above preindustrial levels by the 2080s.[17]

The PIK researchers used the labels *2°C World* and *4°C World* for the two options, and compared them to the baseline period that James Hansen and his associates used, which was 1951–80. The results strongly confirm Hansen's finding that an increase in average temperature leads to a disproportionate increase in heat extremes—going from 2 to 4 degrees more than doubles the extent and magnitude of the damage.

In a 2°C World, unusual (Sigma-3) heat waves will affect about 20 percent of the planet by about 2050 and continue through the century. (The PIK report does not discuss whether this scenario is realistic, but it should be noted that it requires emission cuts much more drastic than any First World government has even suggested, combined with technology that doesn't yet exist.)

In a 4°C World, about 50 percent of the global population, particularly in Africa, Central America, and parts of South America, will frequently experience unusual (Sigma-3) heat waves by 2040—this will be the new summer norm. By 2100, unprecedented (Sigma-5) heat waves will affect 60 percent of the world's land area, particularly the poorest countries where people have the fewest resources for protection, and no way to escape.

This passage from the first *Turn Down the Heat* report is a powerful summary of heat extremes in a 4°C World:

> The effects of 4°C warming will not be evenly distributed around the world, nor would the consequences be simply an extension of those felt at 2°C warming. The largest warming will occur over land and range from 4°C to 10°C. Increases of 6°C or more in average monthly summer temperatures would be expected in large regions of the world, including the Mediterranean, North Africa, the Middle East, and the contiguous United States.
>
> Projections for a 4°C World show a dramatic increase in the intensity and frequency of high-temperature extremes. Recent extreme heat waves such as in Russia in 2010 are likely to become the new normal summer in a 4°C World. Tropical South America, Central Africa, and all tropical islands in the Pacific are likely to regularly experience heat waves of unprecedented magnitude and duration.
>
> In this new high-temperature climate regime, the coolest months are likely to be substantially warmer than the warmest months at the end of the 20th century. In regions such as the Mediterranean, North Africa, the Middle East, and the Tibetan plateau, almost all summer months are likely to be warmer than the most extreme heat waves presently experienced. For

example, the warmest July in the Mediterranean region could be 9°C warmer than today's warmest July.

Extreme heat waves in recent years have had severe impacts, causing heat-related deaths, forest fires, and harvest losses. The impacts of the extreme heat waves projected for a 4°C World have not been evaluated, but they could be expected to vastly exceed the consequences experienced to date and potentially exceed the adaptive capacities of many societies and natural systems.[18]

Bear in mind that the projected 4-degree increase by 2100 is a *mean*, not a maximum. The PIK researchers report that the only models based on RCP 8.5 that include possible warming of less than 4 degrees do so by making extreme assumptions about improved energy efficiency and reduced energy demand.[19] Indeed, according to the latest report from the usually cautious Intergovernmental Panel on Climate Change, if emissions growth continues there is a chance that global average temperature could rise as much as 7.8°C by 2100, a possibility that would make a 4°C World seem entirely benign.[20]

It is virtually certain that our descendants will live in a 4°C World before the century ends, unless greenhouse gas emissions are radically reduced soon. A 4°C World would not be just warmer: almost all the world will be in a *new climate regime*.

That term has a specific meaning in climatology. Moving to a new climate regime means shifting to an environment with a completely different range of climate possibilities. As an extreme example, Antarctica is in a completely different climate regime from Mali or Venezuela—the hottest days at the South Pole are substantially cooler than the coldest days in the tropics. There is no temperature overlap. In the case of a 4°C World, we are talking about a climate regime change in *time* rather than place.

A 2011 paper by scientists at the Stanford University Wood Institute for the Environment examined "the likelihood that rising greenhouse gas concentrations will result in a new, permanent heat regime in which the coolest warm-season of the twenty-first century is hotter than the hottest warm-season of the late twentieth century" if emissions continue to grow.[21] They concluded that by the 2050s, virtually all tropical regions, as well as North Africa and southern Eurasia (which includes the Middle East, South Asia, and Southeast Asia), will be *permanently* in an unprecedented heat regime. By 2070–99, that will also be true in 80 percent of summers in non-tropical North America, China, and the Mediterranean. As the authors point out, this poses major challenges to adaptation:

> In addition to increasing the occurrence of severe hot events, a permanent transition to an unprecedented heat regime could substantially increase climate-related stresses by requiring systems to tolerate a novel temperature envelope in which the new conditions are hotter than the hottest conditions to which those systems are currently accustomed.[22]

Those systems will include the billions of human beings who must live and work in places that will be hotter than anywhere on Earth has been since before our species evolved. The implications for health—and in the longer term for human survival—are frightening. A 2009 study by the British medical journal *The Lancet* concluded that "climate change is the biggest global health threat of the twenty-first century." A follow-up study in 2015 said the threat had grown worse: "The effects of climate change are being felt today, and future projections represent an unacceptably high and potentially catastrophic risk to human health."[23]

The World Health Organization estimates that between 2030 and 2050, climate change will cause millions of additional deaths a year, and the death toll will rise substantially in the second half of the century. The proportion caused by heat stress—directly by extreme heat—will rise fastest of all. [24]

Today, heat stress primarily affects young children and the elderly. As the world warms, workers whose jobs are outdoors or in buildings without air conditioning will be increasingly at risk, when it simply becomes too hot to work safely—at or above the so-called Wet Bulb Globe Temperature (WBGT), at which the human body is no longer able to control its internal temperature.[25]

As a *Lancet* editorial on the 2014 IPCC report said:

> Some scenarios project warming of 4–7°C (on average) over much of the global landmass by the end of the 21st century. If this change happens, then the hottest days will exceed present temperatures by a wide margin and increase the number of people who live in conditions that are so extreme that the ability of the human body to maintain heat balance during physical activity is compromised for parts of the year and unprotected outdoor labor is no longer possible.[26]

By 2100, according to scientists at the U.S. National Oceanic and Atmospheric Administration, "much of the tropics and mid-latitudes [will] experience months of extreme heat stress, such that heat stress in Washington D.C. becomes higher than present-day New Orleans, New Orleans exceeds present-day Bahrain, and Bahrain reaches a WBGT of 31.5°C." The new climate regime, they say, will impose "increasingly severe environmental limitations on individual labor capacity . . . specifically in lost labor capacity in the peak months of heat stress."

> Under both RCP 4.5 and RCP 8.5 by 2050, global lost labor capacity increases in the maximum months to approximately double that in the historical period. Beyond 2050, active mitigation in RCP 4.5 results in reduction of labor capacity to 75% in peak months. Thus, even active mitigation to limit global warming to a 2°C change from pre-industrial conditions results in roughly a doubling of the reduction in lost labor capacity in this model. . . . The highest scenario considered reduces labor capacity to 63% by 2100 in the hottest months.[27]

But not all workers *can* refuse to work, even when their health is in danger. Management pressure to keep going or be fired, economic pressure to avoid losing hourly or piece-work wages—such factors can and will keep production going at the workers' expense. Only militant organization for mutual protection by the workers affected can prevent climate change from becoming a leading cause of death on the job in this century.

Agricultural workers will face a double threat—they won't be physically able to work as long in the fields, and rising temperatures will reduce the crops they depend on for food and income. Recent research has found definite tipping points: when temperatures pass certain levels, crop yields decline rapidly. In a 4°C World, in which local temperature increases of 5 to 10 degrees will affect many areas, the impact on food production may be catastrophic: "Substantial losses are projected for the tropical and subtropical regions and all major crop types. For wheat and maize, losses may even exceed 50 percent on average for large parts of the tropical land area."[28]

In 2015, a panel of experts convened by the United Nations identified some of the major risks posed by 4°C warming:

> In a world 4°C warmer, many global risks are high and very high. Most projected ecosystem impacts would occur at high-risk levels. Climate change velocity would be much too high for terrestrial and freshwater species to move sufficiently fast. Biodiversity losses would take place, including substantial species extinction and disruption of ecosystem services. The risks of combined ocean warming and acidification would become very high. The catch potential of fisheries would be greatly reduced, and crop production would be at very high risk with no potential for adaptation. The long-term sea level rise would far exceed 1 meter, and Arctic summer sea ice would be lost. Some unique systems would be threatened, and the risks from extreme weather events would become very high, or medium to high with adaptation. These risks would put people who are

socially, economically, culturally, politically, institutionally, or otherwise marginalized at most risk. . . .

There are limited prospects for risk reduction through adaptation in a world 4°C warmer, and impacts would significantly increase in all regions. With that amount of warming, limits to adaptation would be reached in relation to aspects such as urban water supply systems, heat-sensitive people, agricultural productivity and food security, means of implementation, and the preservation of cultural identity. Moreover, the adaptation potential in case of conflict over land acquisition and displacement will decline significantly, risks of food insecurity (Africa and Asia) and malnutrition (Africa and Central and South Asia) will be high and very high, and flooding risks will become more widespread in Asia and in Central, South, and North America.[29]

Speeding Down a One-Way Street

And yet, some people tell us not to worry: *"That's just alarmism! You're talking about things that might happen a century from now. We've got lots of time to deal with this."*

Insofar as such "anti-catastrophism" arguments are not just talking points for anti-science lobbyists and their dupes, they reflect deep misunderstanding of the issues. The problem is not that catastrophic change is imminent—it is that unless action is taken soon catastrophic change will become inevitable.

The reason, simply put, is that the amount of global warming depends on the *accumulated volume* of greenhouse gases in the atmosphere—and for over a century, those gases have been entering the atmosphere much more rapidly than natural processes can remove them. Even if all emissions were to stop today, it would be centuries before CO_2 concentrations get back to Holocene levels.

Much of the carbon dioxide that was released into the air by factories in the 1800s is still there, and much of the carbon dioxide being emitted today will still be affecting climate a thousand years from now. The UN's expert panel is blunt: "Anthropogenic climate change, including ocean acidification and many impacts, are *irreversible on at least a multi-century to millennial timescale*."[30] A leading authority on this issue, David Archer of the University of Chicago, is more dramatic, but just as blunt: "The climatic impacts of releasing fossil fuel CO_2 to the atmosphere will last longer than Stonehenge. Longer than time capsules, longer than nuclear waste, far longer than the age of human civilization so far."[31]

Despite many regional fluctuations, the global average temperature in the entire Holocene varied by less than plus or minus 1°C. We are now speeding down a one-way street out of the Holocene: the average global temperature is up by 1°C since 1880, and that small rise has already disrupted the global climate system. There is literally no turning back, and if we continue at the present rate, "by 2100, global average temperatures will probably be 5 to 12 standard deviations above the Holocene temperature mean."[32] That is unprecedented heat, a nightmarish prospect.

It's also important to recognize that forecasts from the IPCC tend to understate the extent and rate of change. In addition to the inherent conservatism of its consensus approach to evaluating risks, four factors tend to make the climate change threat more urgent than official reports suggest.

Tipping points. As we have seen, the Earth System's climate history has been characterized by abrupt changes from one climate regime to another, but such shifts are unpredictable, so even very sophisticated computer models cannot include them: they assume that temperatures will change gradually over time. As global warming develops, there's increasing danger of reaching a tipping point that causes a sudden change in the climate system.

Nine planetary boundaries. Climate change is the most immediate challenge to the stability of the Earth System, but it is only one of the nine planetary boundaries that are in danger. Transgressing multiple boundaries at the same time is likely to multiply their deleterious effect on human development.

2100ism. This term was coined by Australian climatologist James Risbey to describe the tendency of IPCC and other official climate reports "to frame the climate change problem out to the year 2100 . . . [even though] warming and sea level rise will continue well beyond 2100."[33] We saw this around the climate negotiations in Paris in 2015, where discussions of temperature increases simply ignored the time after 2100.

Fat tail. Forecasts of future climate are *probabilities*—Sigma-1 events are described as "likely"—but the probability curve is not symmetrical, it is skewed to the right. There is virtually no chance that the average temperature will rise less than two degrees, and a significant chance that it will rise more. In statistical jargon, the curve has a "fat tail" on the right side. In the book *Climate Shock*, Gernot Wagner and Martin Weitzman show that if the likely range of temperature increase is between 1.5 and 4.5°C, then there is "about a 10 percent chance of eventual temperatures exceeding 6°C."[34] As they point out, the risk that your house will burn down or your car will crash is much less than 10 percent, but you'd be foolish not to insure against those possibilities.

Our rulers give lip service to keeping warming below 2°C, but it is increasingly clear that will not happen if the necessary emission cuts depend on the goodwill of corporations like Exxon and Volkswagen, and on the politicians who serve them. British climatologists Kevin Anderson and Alice Bows-Larkin argue convincingly that keeping the increase below 2°C requires radical action that entails "a reduction in the overall size of the global economy":

> Only if emissions from industrialized nations reduce immediately and at unparalleled rates and only then if less well-off nations begin a rapid transition to low-carbon development with emissions declining from 2025, is there any reasonable probability of not exceeding the 2°C "guard-rail."[35]

As Naomi Klein insists, this should raise the need for radical social change to the top of our agenda:

> What Anderson and Bows-Larkin are really saying is that there is still time to avoid catastrophic warming, but not within the rules of capitalism as they are currently constructed. Which is surely the best argument there has ever been for changing those rules.[36]

FOSSIL CAPITALISM

In April 1856, at a meeting of radical workers in London, Karl Marx described a deep contradiction in capitalist development:

> On the one hand, there have started into life industrial and scientific forces, which no epoch of the former human history had ever suspected. On the other hand, there exist symptoms of decay, far surpassing the horrors recorded of the latter times of the Roman Empire. . . . This antagonism between modern industry and science on the one hand, modern misery and dissolution on the other hand; this antagonism between the productive powers and the social relations of our epoch is a fact, palpable, overwhelming, and not to be controverted.[1]

On another occasion, Marx compared progress under capitalism to that "hideous pagan idol, who would not drink the nectar but from the skulls of the slain."[2] That theme, capitalism's inability to create without destroying, runs like a red thread through Marx's work, but even he could not have imagined how extreme the contradiction would become.

Immense improvements to the human condition have been made in the capitalist era—in health, culture, philosophy, literature, music, and more. But capitalism has also led to starvation, destitution, mass violence, torture, and genocide, all on an unprecedented scale. As capitalism has expanded and aged, the barbarous side of its nature has come ever more to the fore. The great productive forces it creates are, always and simultaneously, immense forces of destruction.

Capitalism and fossil fuels have spectacularly expanded human health and wealth for two centuries. Now they are overwhelming the planetary processes that have made Earth hospitable to civilization and our species for 10,000 years. They are thrusting us into a new and dangerous epoch, the Anthropocene.

The Great Acceleration graphs display twelve critical Earth System trends, from 1750 to the present. Each one is—in itself or as a proxy measurement—a critical component of the Earth System. Each one has operated within clear limits during the 12,000-year Holocene epoch. In their 2015 update to the graphs, climatologist Will Steffen and his colleagues concluded that for nine of the twelve, "there is convincing evidence that the parameters have moved well outside of the Holocene envelope of variability."

1, 2, and 3. The atmospheric concentrations of three greenhouse gases—carbon dioxide, nitrous oxide, and methane—are now well above the maximum observed at any time during the Holocene.
4. There is no evidence of a significant decrease in stratospheric ozone at any time earlier in the Holocene.
5. There is no evidence that human impact on the marine biosphere, as measured by global tonnage of marine fish capture, has been anywhere near the late twentieth-century level at any time earlier in the Holocene.
6. The nitrogen cycle has been massively altered over the past century, and it is now operating well outside of its Holocene range.

7. Ocean acidification is likely changing faster than at any other time in the last 300 million years.
8. Biodiversity loss may be approaching mass extinction rates.
9. Over the 1901–2012 period, global average surface temperature increased by nearly 0.9°C. The global mean temperature from eight to six thousand years ago was about 0.7°C above preindustrial, suggesting that the global climate is now beyond the Holocene envelope of variability. [3]

Humans have always changed the world, but the Anthropocene is something new. The update's authors stress that past changes, no matter how extensive or destructive, did not cause "significant changes in the structure or functioning of the Earth System as a whole."

It is only beyond the mid-20th century that there is clear evidence for fundamental shifts in the state and functioning of the Earth System that are (1) beyond the range of variability of the Holocene, and (2) driven by human activities and not by natural variability.

They conclude: "We have reached a point where many biophysical indicators have clearly moved beyond the bounds of Holocene variability. We are now living in a no-analog world."[4] There is now no doubt that global climate is outside the Holocene envelope of variability. We are in *terra incognita*, the uncharted territory known as the Anthropocene.

Part One discussed the Anthropocene as a *biophysical* phenomenon—a qualitative change in Earth's most critical physical characteristics that has profound implications for all living things, including humans. That's important, but to properly understand the Anthropocene we must see it as a *socio-ecological* phenomenon—a qualitative change in the relationship between human society and the rest of the natural world. It is a direct result, to use Marx's phrase, of the "irreparable rift in the interdependent process of social metabolism, a metabolism prescribed by the nat-ural laws of life itself."[5]

Part Two examines why and how two hundred years of capitalist development brought an end to the Holocene and began "driving the Earth System onto a trajectory toward more hostile states from which we cannot easily return."[6]

The search for the social and economic origins of the Anthropocene is very different from the search for its geological beginning. Geology, by its nature, looks for a clear physical tran-sition in rock, sediment, or ice, a place where a "golden spike" (actually a brass marker) can be placed to formally demark one epoch from the next. Social science cannot be so precise. As Lenin wrote in his famous account of the rise of imperialism, "All bound-aries in nature and in society are conventional and changeable, and it would be absurd to argue, for example, about the particular year or decade in which imperialism 'definitely' became established."[7]

The shifts shown in the Great Acceleration graphs were not pro-duced by a single event, such as a collision with a comet: they were the culmination of two centuries of capitalist development. So while geologists search for an exact decade or even an exact day, a Marxist analysis looks for a longer period of social and economic change during which the Holocene ended and the Anthropocene began.

7

Capital's Time vs. Nature's Time

The modern world worships the gods of speed and quantity, and
of quick and easy profit, and out of this idolatry monstrous evils
have arisen.

—RACHEL CARSON[1]

In article after article, book after book, scientists and environmentalists have exposed the devastating effects of constant
economic expansion on the global environment. The drive to
produce ever more "stuff" is filling our rivers with poison and our
air with climate-changing gases. The oceans are dying, species are
disappearing at unprecedented rates, water is running short, and
soil is eroding much faster than it can be replaced.

But the growth machine pushes on. Corporate executives, economists, pundits, bureaucrats, and of course politicians, all agree
that growth is good and non-growth is bad. Unending material
expansion is a deliberate policy promoted by ideologues of every
political stripe, from social democrats to ultraconservatives. When
the leaders of the world's richest countries, the G20, met in Toronto
in 2010 they unanimously agreed that their highest priority was to
"lay the foundation for strong, sustainable, and balanced growth."
The word *growth* appeared 29 times in their final declaration.[2]

Why, in the face of massive evidence that constant expansion of production and extraction of resources is killing us, do

governments and corporations keep shoveling coal for the runaway growth train?

In most environmental writing, one of two explanations is offered—it's human nature, or it's a mistake.

The human nature argument is central to mainstream economics. Our species is *homo economicus*, economic man, defined by John Stuart Mill as "a being who inevitably does that by which he may obtain the greatest amount of necessaries, conveniences, and luxuries, with the smallest quantity of labor and physical self-denial with which they can be obtained."[3] So we always want more, and economic growth is just capitalism's way of meeting that fundamental human desire. For our species, enough is never enough.

That view often leads its proponents to conclude that the only way to slow or reverse the pillaging of Mother Earth is to slow or reverse population growth. More people equals more stuff; fewer people equals less stuff. As Simon Butler and I showed in our book *Too Many People?*, many populationist arguments are no more sophisticated than that. The fact that the countries with the highest birth rates generally have the lowest standard of living and produce the least pollution fatally undermines such claims—if the poorest 3 billion people on the planet somehow disappeared tomorrow, there would be virtually no reduction in ongoing environmental destruction.

Seduced By a False Ideology?

The other common greenish explanation for the constant promotion of growth is that we have been seduced by a false ideology. For example, environmental science professor Robert Nadeau argues that political leaders and economic planners are under the sway of "a quasi-religious belief system," so what is needed is a religious conversion. "If political leaders and economic planners realize that the gods they now serve are false and proceed to do what is required to resolve the environmental crisis, we can soon be living in a very different world."[4] Other writers describe the drive for growth as a *fetish*, an *obsession*, an *addiction*, or even a *spell*.

Technology writer Fawzi Ibrahim points out that such accounts present capitalism's inexorable drive to expand not as an inevitable result of the profit system, but as "a psychological obsession, a sort of collective whim that simultaneously grips all governments and all economists worldwide, or some kind of elaborate global conspiracy."[5]

> The importance given to growth . . . is seen as a stand-alone phenomenon, a matter of individual choice by economists or a collective decision by government and society that can be turned on and off. It must be the first time in history that a necessity has been described as a fetish. You might as well describe fish having a fetish for water as capitalism having a fetish for growth. Growth is as essential to capitalism as is water to fish. As fish would die without water, so would capitalism drown without growth.[6]

For millennia almost all production was for use, so there was little need or room for economic growth as we understand it today. But under capitalism, most production is for exchange: capital exploits labor and nature to produce goods that can be sold for more than the cost of production, in order to accumulate more capital, and the process repeats. Growth ideology doesn't *cause* perpetual accumulation—it *justifies* it.

Capital's Prime Directive

If you were to ask them individually, the people who run Exxon, Volkswagen, and other giant polluters would undoubtedly tell you that they want their children and grandchildren to live in a clean, environmentally sustainable world. But as major shareholders and executives and top managers they act, in Marx's phrase, as "personifications of capital." Regardless of how they behave at home or with their children, at work they are capital in human form, and the imperatives of capital take precedence over all other needs and values. So when it comes to a choice between protecting humanity's future and maximizing profit, they choose profit.

A person who is unwilling to put the needs of capital first is not likely to reach the executive suite of a major corporation. If the screening process fails, or if a corporate executive has an inconvenient attack of conscience, he or she will not last long in that position. It has been called the *ecological tyranny of the bottom line*: when protecting humanity and planet might reduce profits, corporations will always put profits first.

Capital's only measure of success is *accumulation*. How much more profit was made in this quarter than in the previous quarter? How much more today than yesterday? It doesn't matter if the sales include products that are directly harmful to both humans and nature, or that many commodities cannot be produced without spreading disease, destroying the forests that produce the oxygen we breathe, demolishing ecosystems, and treating our water, air, and soil as sewers for the disposal of industrial waste. It all contributes to the growth of capital—and that is what counts.

Capital has existed in various forms for thousands of years, but only in the past five centuries has it defined and dominated the economy as a whole. A small minority owns almost all of the profit-producing wealth and lives on the proceeds, while most people must work for that minority in order to survive. Employees produce more wealth than they are paid, and capital grows by taking part or all of the surplus.

Each corporation seeks to ensure that its products are sold at prices sufficient to produce an attractive profit on invested capital. A corporation with lower costs or more attractive products can drive its competitors out of business. This creates irresistible pressure on all corporations to improve productivity, reduce costs, or otherwise make their products more saleable—and that pushes the system as a whole to expand physically, financially, and geographically. As Marx and Engels wrote in *The Communist Manifesto*:

> Constant revolutionizing of production, uninterrupted disturbance of all social conditions, everlasting uncertainty, and agitation distinguish the bourgeois epoch from all earlier ones. All fixed, fast-frozen relations, with their train of ancient and venerable prejudices and opinions, are swept away, all new-formed ones become antiquated before they can ossify. All that is solid melts into air, all that is holy is profaned.[7]

If nothing stops it, capital will try to expand infinitely—but Earth is not infinite. The atmosphere and oceans and the forests are very large, but ultimately they are finite, limited resources—and capitalism is now pressing against those limits. Greenhouse gas emissions are not unusual or exceptional. Pouring crap into the environment is a fundamental feature of capitalism, and it is not going to stop so long as capitalism survives. That is why "solutions" like cap-and-trade have failed so badly and will continue to fail: waste and pollution and ecological destruction are built into the system's DNA.

In *Seventeen Contradictions*, David Harvey asks whether capital accumulation might not level off, leading to a zero-growth, steady-state capitalist economy—and the answer he comes to is "a resounding no."

> The simplest reason is that capital is about profit seeking. For all capitalists to realize a positive profit requires the existence of more value at the end of the day than there was at the beginning. That means an expansion of the total output of social labor. Without that expansion there can be no capital. A zero-growth capitalist economy is a logical and exclusionary contradiction. It simply cannot exist.[8]

Of course, the fact that capital *needs* to grow does not mean that it always *can* grow. On the contrary, the drive to grow periodically leads to situations in which more commodities are produced than can be sold: the result is a crisis in which substantial amounts of wealth are destroyed. Individual corporations can and do go out of business in such situations, but over the long term the drive for profit, to accumulate ever more capital, always reasserts itself: it is a defining feature of the capitalist system and the root cause of the global environmental crisis.

"A Metabolism Prescribed by the Natural Laws of Life"

In the mid-1800s, the German scientist Justus von Liebig explained the decline of agricultural productivity in England by showing that in its natural state, soil provided essential nutrients that allowed plants to grow, and replenished its store of those nutrients from plant and animal waste. Capitalist agriculture interrupted that cycle by preventing the return of wastes, causing soil fertility to decline. Liebig used the then-new word *metabolism* (*Stoffwechsel*) for the interactions and chemical/biological exchanges between plants, animals, and soil.

Since then, the concept of metabolism has been "a key category in the systems theory approach to the interaction of organisms to their environments."

> It captures the complex biochemical process of metabolic exchange, through which an organism (or a given cell) draws upon materials and energy from its environment and converts these by way of various metabolic reactions into the building blocks of growth. In addition, the concept of metabolism is used to refer to the specific *regulatory processes* that govern this complex interchange between organisms and their environment.[9]

Liebig's work laid the ground for a new approach to the natural world, one that viewed it not as an assemblage of separate things—plants, animals, humans, rivers, atmosphere, etc.—but as systems in which all components constantly interact and in which any change to any part can change others. In the 1930s, ecologist

Arthur Tansley, a Fabian socialist, coined the word *ecosystem* for biological communities, "including not only the organism-complex, but also the whole complex of physical factors forming what we call the environment."[10]

A SYSTEM THAT NEVER STANDS STILL

The purpose of capitalist enterprise has always been to maximize profit, never to serve social ends. Mainstream economic theory since Adam Smith has insisted that by directly maximizing profit the capitalist (or entrepreneur) is indirectly serving the community. All the capitalists together, maximizing their individual profits, produce what the community needs while keeping each other in check by their mutual competition. All this is true, but it is far from being the whole story. Capitalists do not confine their activities to producing the food, clothing, shelter, and amenities society needs for its existence and reproduction. In their single-minded pursuit of profit, in which none can refuse to join on pain of elimination, capitalists are driven to accumulate ever more capital, and this becomes both their subjective goal and the motor force of the entire economic system.

It is this obsession with capital accumulation that distinguishes capitalism from the simple system for satisfying human needs it is portrayed as in mainstream economic theory. And a system driven by capital accumulation is one that never stands still, one that is forever changing, adopting new and discarding old methods of production and distribution, opening up new territories, subjecting to its purposes societies too weak to protect themselves. Caught up in this process of relentless innovation and expansion, the system runs roughshod even over its own beneficiaries if they get in its way or fall by the roadside. As far as the natural environment is concerned, capitalism perceives it not as something to be cherished and enjoyed but as a means to the paramount ends of profit-making and still more capital accumulation.

—PAUL M. SWEEZY[11]

As we have seen, in recent decades scientists have gone beyond ecology to study Earth itself as an integrated system of systems. One important result has been a much clearer—and much more worrying—understanding of the system-wide relationship between human society and the rest of the planet.

To understand why a system that has operated smoothly for thousands of years is now being seriously disrupted, it is useful to consider the ecological insights of a nineteenth-century radical who studied Liebig's work on the metabolism of agricultural soil very carefully. That was Karl Marx.

One of the most important books of Marxist theory published in recent years is John Bellamy Foster's *Marx's Ecology: Materialism and Nature*. In it, Foster rediscovered and expanded on a long-ignored feature of Marx's work, the concept of *metabolic rift*. The idea had been hiding in plain sight in the pages of *Capital* for over a century, but not until the global environmental crises of the late twentieth century did most socialists become reacquainted with Marx's powerful insights into the dysfunctional relationship between capitalism and the natural world it depends upon.

Marx studied Liebig's work carefully, telling his lifelong comrade Frederick Engels in 1866 that "the new agricultural chemistry . . . is more important for this matter than all the economists put together."[12] He seized upon the concept of *metabolism*, of material cycles that are essential to life, and made it central to his analysis of the relationships between humanity and nature. In *Capital*, Marx integrated Liebig's explanation of the chemistry of the soil exhaustion crisis into his historical and social analysis of capitalism, showing how the imperatives of capitalist growth inevitably conflict with the laws of nature.

> Capitalist production collects the population together in great centers, and causes the urban population to achieve an ever-growing preponderance. This has two results. On the one hand it concentrates the historical motive force of society; on the other hand, it disturbs the metabolic interaction between man

and the earth, i.e. it prevents the return to the soil of its con-
stituent elements consumed by man in the form of food and
clothing; hence it hinders the operation of the eternal natural
condition for the lasting fertility of the soil.[13]

Instead of growing food for use, capitalist agriculture grows it
for sale and profit: the products of Earth are sent to the cities, but
humanity's wastes are not returned to the land. Essential nutrients dumped elsewhere become pollutants: the capitalist economy,
Marx wrote, can "do nothing better with the excrement produced
by 4½ million people than pollute the Thames with it at monstrous
expense."[14]

Marx described this as "an irreparable rift in the interdependent
process of social metabolism, a metabolism prescribed by the natural laws of life itself."[15] The concept of a metabolic rift expresses
humanity's simultaneous dependence on and separation from the
rest of nature. Foster explains:

> Marx employed the concept of a "rift" in the metabolic relation
> between human beings and the earth to capture the mate-
> rial estrangement of human beings within capitalist society
> from the natural conditions which formed the basis for their
> existence—what he called "the everlasting nature-imposed
> condition[s] of human existence."[16]

Australian socialist and environmental activist Del Weston
explained the concept of metabolic rift particularly well in her
book, *The Political Economy of Global Warming*. She described metabolic rift as "the crux of Marx's ecological critique of capitalism":

> In a general sense, the metabolic rift refers to a disruption
> between social systems and natural systems, leading to eco-
> logical crisis. In Marxist ecological theory, humans exist in a
> "metabolic" relation with nature that is fundamental for sur-
> vival. The metabolic relation is people's labour, the material

process by which humans transform the raw materials of nature to meet their material needs. . . .

[It] is at the crux of Marx's ecological critique of capitalism, denoting the disjuncture between social systems and the rest of nature. It is an inherent part of the capitalist social relations of production and is found, for example, in expressions of capitalism's relationship to soil fertility, in the relationship of capital to labour, and in the imperialist nature of capitalism. . . .

The metabolic rift thus refers to a rupture in the metabolism of the whole ecological system, including humans' part in that system. The concept is built around how the logic of accumulation severs basic processes of natural reproduction, leading to the deterioration of the environment and ecological sustainability and disrupting the basic operations of nature. It neatly captures the lack of balance between "expenditure and income" in the Earth's metabolism under the capitalist system.[17]

An Incurably Short-Term Horizon

Capital's ecologically destructive impacts result not just from its need to grow, but from its need to *grow faster*. The circuit from investment to profit to reinvestment requires time to complete, and the longer it takes, the less total return investors receive. Other things being equal, an investment that returns the expected profit sooner will attract more capital than one that pays off later. If I get my money back more quickly, I can "put it to work again" more quickly: two circuits in a year is better than one, three is better than two, and so on. Competition for investment produces constant pressure to speed up the cycle, to go from investment to production to sale ever more quickly. As Rachel Carson wrote, in the modern world, "the rapidity of change and the speed with which new situations are created follow the impetuous and heedless pace of man rather than the deliberate pace of nature."[18]

That is why it took sixteen weeks to raise a two-and-a-half-pound chicken in 1925, while today it takes just six weeks to raise

---

"THE MARKET SYSTEM IS WORKING TOO WELL"

The essential problem for the Earth—for us—is that there is a mismatch between the short timescales of markets and the political systems tied to them, and the much longer timescales that the Earth System needs to accommodate human activity. The climate crisis is upon us not because markets aren't working well enough but because the market system is working too well in accelerating global energy and material cycles. Technological progress and the globalization of finance, transport and communications have oiled the wheels of the human-willed components of the planetary system, allowing them to accelerate. Put another way, the tempo of the market's metabolism is much faster than that of the Earth System, yet in the Anthropocene they no longer operate independently.

—CLIVE HAMILTON[19]

---

one twice as big. Selective breeding, hormones, and chemical feed have enabled factory farms to produce not just more meat, but *more meat faster*. The suffering of the animals and the quality of the food are secondary concerns, if they are considered at all.

Most natural processes cannot be manipulated that way. As Marx wrote of farming, "The entire spirit of capitalist production—which is oriented toward the most immediate monetary profit—stands in contradiction to agriculture, which has to concern itself with the whole gamut of permanent conditions of life required by the chain of human generations."[20] In its endless pursuit of profit, capital destroys the soil, even though that will deprive future generations of food. Timber industries wiped out the great forests that once covered Europe, and never replaced them, because "the long production time . . . and the consequent length of the turnover period, makes forest culture a line of business unsuited to private and hence to capitalist production."[21]

Nature's cycles operate at speeds that have evolved over many millennia—forcing them in any way inevitably destabilizes the cycle and produces unpleasant results. Fertile land is destroyed, forests are clear-cut, and fish populations collapse, all because capitalism needs to operate at speeds much faster than the natural cycles of reproduction and growth. Christopher Wright and Daniel Nyberg justly describe such irrationality as "processes of creative self-destruction":

> Our economic system is now engaged in ever more inventive ways to consume the very life-support systems upon which we rely as a species; moreover, this irrational activity is reinvented as a perfectly normal and sensible process to which we all contribute and from which we all benefit.[22]

Liberal greens sometimes suggest that eliminating quarterly financial reports would encourage corporate executives to think longer-term and act more responsibly toward nature. Such naïve proposals confuse cause and effect, ignoring what Mészáros calls the *incurably short-term horizon of the capital system*. "It cannot be other than that, in view of the derailing pressures of competition and monopoly and the ensuing ways of imposing domination and subordination, in the interest of immediate gain."[23]

There is, in short, an insuperable conflict between nature's time and capital's time—between the cyclical Earth System processes that have developed over millions of years, and capital's need for rapid production, delivery, and profit.

Global Metabolic Rifts: Carbon and Nitrogen

The metabolic processes that Liebig and Marx knew of and wrote about were local or regional: a break in metabolic processes on one farm did not necessarily affect neighboring farms. Colonialism extended the damage by transporting chemicals and products from distant places. Describing how England imported food from poverty-stricken Ireland, Marx wrote: "England has indirectly exported the soil of Ireland, without even allowing its cultivators the means for replacing the constituents of the exhausted soil."[24] But even in such cases, the areas affected were limited.

Only in the 1970s, with the discovery that CFCs were damaging the ozone layer, did it become evident that what seemed to be routine economic activity could rupture natural processes that are essential to the continued operation of the planet as a whole. That realization contributed directly to the birth of Earth System science, and that in turn has led to much better understanding of the metabolic processes that shape the Earth System. The Planetary Boundaries Framework, discussed in chapter 4, attempts to define the natural processes and systems that regulate the stability and resilience of the Earth System—and the boundaries themselves attempt to define limits beyond which those processes and systems can no longer function as they have done for at least 11,000 years. The individual boundaries are current and potential global metabolic rifts.

In a 1969 article, ecologist Barry Commoner used poetic language to describe four key elements that are essential to all life:

> Four chemical elements make up the bulk of living matter—carbon, hydrogen, oxygen, and nitrogen—and they move in great, interwoven cycles in the surface layers of the earth: now a component of the air or water, now a constituent of a living organism, now part of some waste product, after a time perhaps built into mineral deposits or fossil remains.[25]

Global metabolic rifts have now developed in two of those four chemical cycles.

The **Carbon Cycle** regulates the energy balance of the climate system. Atmospheric carbon dioxide (CO_2) is transparent to visible sunlight, but opaque to infrared heat energy, so it lets light in and hinders heat from escaping into space, which means that an increase in CO_2 causes warming, and a decrease causes cooling. Under natural conditions, CO_2 enters the atmosphere from deep within the Earth through volcanoes and hot springs, and returns to the Earth through chemical reactions, when CO_2-laden water weathers rock. Weathering is faster when temperatures are high, slower when they are low—so in warm periods it removes CO_2 more quickly and temperatures fall, while in cold periods it removes less CO_2 and temperatures rise. This process stabilizes climate, but only in the very long term—it can take hundreds of thousands or even millions of years to return to equilibrium after a substantial CO_2 increase.

Fossil fuels have disrupted that cycle. Over 300 million years ago, long before dinosaurs arrived, geological processes buried trees, ferns and other plants deep in the earth, removing them from the global carbon cycle. Now that buried carbon is oil, gas, and coal, and is being burned in a fraction of a geological second. What nature took hundreds of millions of years to create, capital is destroying in a few hundred years, releasing CO_2 into the atmosphere much faster than weathering and other natural processes can remove it. The global temperature is rising, and, as we saw in chapter 6, that is changing the state of the entire planet.

A less visible but just as serious metabolic rift affects the **Nitrogen Cycle**. All living things need nitrogen: plants cannot grow without it, and animal bodies (including ours) need it to make muscles, skin, blood, hair, nails, and DNA. Traditionally, farmers maintained nitrogen levels in the soil by crop rotation and by adding animal manure to the soil, but the intensive, market-oriented agriculture that began in the 1800s depleted the soil faster than natural processes could regenerate it. Fertilizer based on bird guano, which contains nitrogen, or on nitrate mineral deposits helped for a while, but those sources were running out by the 1890s. In 1909, German chemists discovered a way—a very energy-intensive way—to extract nitrogen from the atmosphere. Today, the Haber-Bosch process produces more usable nitrogen than all natural processes combined.

Over-fertilizing, either accidental or deliberate, to maximize crop growth is common: the excessive nitrogen produces a host of environmental problems, including coastal dead zones and fish kills; biodiversity loss; pollution of lakes, rivers, and groundwater; respiratory diseases; global warming and ozone depletion. A bitter irony is that excessive use of nitrogen fertilizer reduces the soil's fertility, so more and more must be used to maintain production.

The Planetary Boundaries Framework recommends reducing the atmospheric CO_2 level to less than 350 parts per million (it is now over 400) and reducing artificial nitrogen production to less than 25 percent of natural production.

In each case, capital's time is overwhelming nature's time, producing rifts that are big enough to be called metabolic chasms. In Del Weston's words, the metabolic rift "has grown both in dimensions and complexity, to the point where the economic activities of human society are causing an unprecedented change in the Earth's biosphere, its lands, forests, water, and air, potentially bringing to an end the Holocene era as a result of anthropogenic global warming."[26]

The question, then, is how did this happen?

8

The Making of Fossil Capitalism

> Another basic contradiction of the capitalist system of control is that it cannot separate "advance" from destruction, nor "progress" from waste—however catastrophic the results. The more it unlocks the powers of productivity, the more it must unleash the powers of destruction; and the more it extends the volume of production, the more it must bury everything under mountains of suffocating waste.
>
> —ISTVÁN MÉSZÁROS[1]

The preceding chapter discussed the anti-ecological characteristics of capitalism in general—its fundamental drive to grow and speed up, and its consequent tendency to override and rupture nature's essential processes and cycles. Those factors underlie today's Earth System crisis, but they don't explain its particular character. Marx and Engels rejected attempts to deduce social and political developments from first principles:

> Our view of history . . . is first and foremost a guide to study, not a tool for constructing objects after the Hegelian model. The whole of history must be studied anew, and the existential conditions of the various social formations individually investigated before an attempt is made to deduce therefrom the

political, legal, aesthetic, philosophical, religious, etc., stand-
points that correspond to them.[2]

Recognizing that capitalism causes environmental destruction
is the beginning of ecological wisdom, but only the beginning. As
Ernest Mandel wrote in a different context, every capitalist crisis
has "combined general features related to the fundamental contradictions of the capitalist mode of production with particular
features resulting from the precise historical moment of the development of that mode of production in which it occurs."[3] The Great
Acceleration and Earth's transition to the Anthropocene are not
exceptions. Capitalism has been wrecking ecosystems for hundreds of years, but its assault on the entire Earth System at once
is a recent development. That unique transformation requires
explanation.

In the eighteenth and nineteenth centuries, capital's need to grow
drove an epochal shift to a fossil fuel–based economy, becoming,
in Engels's telling phrase, a "squanderer of past solar heat."[4] Using
solar energy that had been preserved underground for millions
of years allowed production to break through the limitations of
wind, water, and muscle. The cost of obtaining and using fossil fuel
was vastly less than the resulting profits. As we know now, there
was a huge hidden cost in environmental destruction, especially
the disruption of the carbon cycle, but such costs don't factor into
capital's accounting.

Burning fossil fuels pumped carbon dioxide and other gases
into the atmosphere, leading to a disruption of the carbon cycle
that manifests as acidification of the oceans and climate change.
The rift in Earth's carbon metabolism widened slowly for a century, and then reached a tipping point in the years following the
Second World War. The Anthropocene, which in retrospect had
been a possibility since the Industrial Revolution began, became
a reality in the second half of the twentieth century, when the
rift in the carbon cycle suddenly expanded past the point of no
return.

Coal, Steam, and Capital

Western European capitalism was born in the fifteenth and six-teenth centuries, but it didn't spring into being fully formed. In Marx's view, "a specifically capitalist mode of production" requires "large-scale industry" that "not only transforms the situations of the various agents of production, it also *revolutionizes* their actual mode of labor and the real nature of the labor process as a whole."[5] That transformation began in England in the Industrial Revolution, and since then the history of the specifically capitalist mode of production has been inseparable from the history of fossil fuels.

When capitalism first arose in the 1400s, the principal sources of energy were wood, wind, water, and human or animal muscle. Coal was used on a limited scale, but it didn't become a significant factor in production until the 1700s, when it began to replace wood in production processes that required heat—brewing and soap-making, for example—and for heating homes and cooking.

In its raw form, impurities in coal made it unsuitable for smelting iron ore, so that industry continued to use charcoal until well into the 1700s. Producing charcoal used so much wood that serious fuel shortages developed, and Britain had to import iron from Sweden. That crisis was overcome late in the century when the iron industry switched from charcoal to coke—coal that had been cooked to remove impurities. That shift drove a huge increase in iron production and a parallel increase in coal mining. Readily available iron in turn enabled large-scale production of factory machinery and steam engines, and, in the 1800s, made the rail-road boom possible. The key elements of the Industrial Revolution all depended on coal.

As Andreas Malm's insightful account in *Fossil Capital* shows, general adoption of coal and steam for manufacturing was far from an automatic process. At the beginning of the 1800s, most of the cotton spinning and weaving factories that are virtu-ally synonymous with the early factory system were powered by river-driven water wheels, not coal. In 1800, a full quarter century after James Watt perfected his machine, only 84 English cotton mills were using steam engines, compared to a thousand that used water wheels. The number of steam-powered mills didn't equal the number using water power until about 1830. Malm demonstrates that, contrary to the conventional narrative, the shift to coal and steam did not occur because it was cheaper or more reliable—it was neither—but because it gave the factory owners better access to and control over labor. Water-powered mills had to be located by rapids or waterfalls, usually in rural areas where the pool of potential workers was small; steam-powered mills could be located in towns and cities, close to large numbers of workers who were used to factory work, and where the presence of an army of the unemployed made it easier to replace any workers who didn't meet the owners' requirements.[6]

While mill owners were adopting coal and steam as the energy basis of production, other capitalists were doing the same for transportation. Steam locomotives, which are essentially steam engines on wheels, were used to haul coal from mines to canals or seaports as early as 1804, and the first public railroad in England, built to carry raw cotton from the port of Liverpool to cotton mills in Manchester, began service in 1830. Its success—investors received a 9.5 percent annual dividend for fifteen years—triggered a boom in railroad building, including a speculative bubble in the 1840s in which thousands of investors lost everything. By the 1850s, some 6,200 miles of track had been laid in Britain.[7] In following years, railroads grew particularly rapidly on the other side of the Atlantic: by 1877 there were over 79,000 miles of track in the United States, and one railroad company, the Pennsylvania, was by far the largest corporation in that country.[8]

Britain was a generation ahead of other capitalist countries in adoption of a fossil fuel regime: in 1825 it was producing 80 percent of the world's greenhouse gas emissions from fossil fuel combustion, and in 1850 it was still generating 62 percent—twice as much as the United States, France, Germany and Belgium combined.[9]

Between 1850 and 1873, Britain's coal consumption tripled, from 37 million to 112 million tons; France's jumped from 7 million to nearly 25 million tons; and Germany's rose from 5 million to 36 million.[10]

But by the end of the century, industry and railroads in the United States were burning more coal than those in Britain, Germany was about to catch up, and several other European countries were getting close.

Fuel, Empire, and War

As Bruce Pobodnik has shown, "The global energy shift toward coal that occurred during the early nineteenth century not only transformed societies within Europe, but it also had far-reaching global consequences":

> This energy shift became intimately associated with a new process of conquest that forcibly incorporated new regions into an expanding world-system. Coal-powered ships and railroads allowed Britain and its Continental rivals to seize control over territories in Asia, Africa, and the Middle East that had long resisted conquest.[11]

Armed steamboats patrolled India's rivers, and the British navy used steam-powered gunboats to defeat China's much larger sail-based fleet in the so-called Opium Wars of the 1840s and 1850s. Britain, France, and Germany all used steamboats in colonial wars, "fundamentally shifting the balance of military power between Europe and even the strongest Asian and African societies."[12] In the late 1800s, a key assignment for British naval expeditions to Africa, Asia, and the Pacific was to find and take control of coal deposits and establish coaling stations so that naval and merchant ships could travel the world unhindered.

Railroads played important roles in moving troops and munitions in the U.S. Civil War and the Franco-Prussian War of

1870–71, leading governments in Europe and North America to subsidize coal mining and railway building with cash, land grants, and, large military contracts. "The period from 1880 on," writes Pobodnik, "marks the emergence of the first true military-industrial complexes, with European, North American, and even Japanese companies entering into long-term development projects with army and naval contractors."[13]

The invention of the internal combustion engine in the 1880s, and of the airplane in 1903, created a new market for petroleum, hitherto used mainly for lighting (as kerosene) and lubrication. The new machines used gasoline, a part of petroleum that refiners had been discarding as useless and dangerous. In the first two decades of the twentieth century imperial armies became major customers for gasoline. Italy used airplanes against Turkey in 1911, and against rebels in Morocco in 1912, and the United States used nearly six hundred trucks in its 1916 attack on Pancho Villa's rebel forces in Mexico.

But the big military breakthrough for petroleum as a fuel was Britain's decision, in 1912, to convert its battleships from coal to oil. As with the switch from water to coal in British cotton mills, class struggle played a major role in the decision. In 1910, Winston Churchill had used the army to break strikes in the Welsh coal mines that were the only source of the high-grade anthracite coal required by battleships. When he assumed responsibility for the navy in 1911, he immediately initiated a program to convert the battleships to oil. "In committing the Royal Navy to a new source of energy, the government . . . was freeing itself from the political claims of the coal miners."[14]

Oil-powered tanks, airplanes, destroyers, and submarines played decisive roles in the barbaric slaughter known as World War I. Britain alone deployed nearly 100,000 trucks and cars, while the United States used some 50,000 vehicles and 15,000 airplanes. Between 1914 and 1918, the global tonnage of oil-powered ships more than tripled, and production of gasoline-powered vehicles increased five-fold.[15]

Automobilization

In 1966, in the pathbreaking book *Monopoly Capital*, Paul Baran and Paul Sweezy discussed "epoch-making innovations . . . [that] shake up the entire pattern of the economy and hence create vast investment outlets in addition to the capital which they directly absorb. . . . Only three really meet the 'epoch-making' tests: the steam engine, the railroad, and the automobile."[16]

All three depended on fossil fuels, so it may be more accurate to describe the epoch-making innovations as coal/steam engine, coal/railroad, and oil/automobile. These combinations of technology and fuel brought the specifically capitalist mode of production to full maturity as a fossil economy.

The first automobiles were individually built and sold as expensive toys for the very rich, so initially the gasoline market was small. That changed quickly when the automobile industry, led by the Ford Motor Company, adopted mass-production techniques and technology. As with textiles in Britain in the 1800s, the shift to a specifically capitalist mode of production was revolutionary. Costs and prices fell, and sales soared.

The first phase of the process that Baran and Sweezy called "automobilization" occurred from just before World War I to about 1929 in the United States. Sales of cars, trucks, and buses soared from 4,000 in 1900 to 1.9 million in 1919 to 5.34 million in 1929, by which time automobile manufacturing was the country's largest industry. In addition to selling millions of cars, the industry itself was a huge market for steel, glass, and rubber.

Most important, for this discussion, "automobiles and trucks transformed the petroleum industry from a producer of illuminants and lubricants into a supplier of gasoline."[17] The oil companies built an unprecedented sales and support network. In 1920, gasoline was an anonymous product sold as a sideline by hardware and grocery stores, but by 1929 there were over 120,000 brand-name filling stations across the country and the number doubled by 1939.[18] Two-thirds of the world's oil came from wells in the United States, and U.S. oil companies ranked among the biggest and most profitable corporations in the world.

Industrial Chemistry

While the oil companies were building markets for petroleum-as-fuel, the chemical industry was developing entirely new products that were either made from the by-products of oil-refining or that required high levels of energy that only oil could provide, or both.

The key technologies of the Industrial Revolution—steam engine, spinning jenny, cotton gin, and so on—were invented by individual tinkerers working alone or with a few assistants. As late as 1903, the first heavier-than-air flying machine was designed and built by two brothers in a bicycle shop. But by then, invention itself had become big business: corporations were hiring scientists and creating wholly-owned research laboratories to invent on demand. There were 300 corporate laboratories in the United States in 1920, and over 2,200 in 1940.[19] On the eve of World War II, thirteen U.S. companies employed one-third of all research scientists in the country.[20] In chapter 5 one important product of this merger of science and big business was discussed, the 1930 introduction of chlorofluorocarbons (CFCs) for refrigeration, air conditioning, and aerosol sprays by the General Motors/DuPont group.

Ernest Mandel describes this development as part of a third technological revolution, "an epoch of unprecedented fusion of science, technology, and production,"[21] and Harry Braverman calls it the scientific-technological revolution:

> The contrast between science as a generalized social property incidental to production and science as capitalist property at the very center of production is the contrast between the Industrial Revolution, which occupied the last half of the eighteenth and the first third of the nineteenth centuries, and the scientific-technical revolution, which began in the last decades of the nineteenth century and is still going on.[22]

Before World War I, the U.S. chemical industry mainly manufactured German-developed products, under license from companies such as BASF and Bayer, the undisputed world leaders in industrial chemistry, but one week before the war ended the United States confiscated all German patents under the Trading with the Enemy Act and began licensing them at low cost to American firms. This laid the basis for the rapid growth of U.S. chemical companies such as DuPont, American Cyanamid, Dow Chemical, and Monsanto. The 1930s saw the invention of artificial fibers such as nylon and rayon, and the first mass-produced plastics, as well as a wide range of new industrial chemicals. The first new synthetics were made from coal tar, but by the end of the 1930s, the industry had largely converted to natural gas and by-products of petroleum refining as feedstock for new products, and the giant oil companies were playing a major role in what came to be called the petrochemical industry.[23]

Concentrated Capital

In 1930, 106 of the 200 largest industrial companies in the United States were in the chemicals, petroleum, metals, rubber, or transportation industries, all closely related to the petroleum-automobile complex.[24] This extraordinary growth and concentration was part of a long-term trend: since the 1880s, capitalism had been rapidly shifting from a system in which every industry comprised many small- and mid-sized firms to one dominated by a small number of gigantic companies. In 1916, Lenin calculated that 1 percent of enterprises in the United States was responsible for almost half of all production, and that less than 1 percent of companies in Germany used more than three-quarters of all steam and electric power.[25] According to the Federal Trade Commission, the 200 largest U.S. companies had 35 percent of the turnover of all companies in 1935, 37 percent in 1947, 40.5 percent in 1950, and 47 percent in 1958.[26]

As Paul Baran and Paul Sweezy wrote in *Monopoly Capital*, the result was a different kind of capitalism:

Competition, which was the predominant form of market relations in nineteenth-century Britain, has ceased to occupy that position, not only in Britain but everywhere else in the capitalist world. Today the typical economic unit in the capitalist world is not the small firm producing a negligible fraction of a homogeneous output for an anonymous market but a large-scale enterprise producing a significant share of the output of an industry, or even several industries, and able to control its prices, the volume of its production, and the types and amounts of its investments.[27]

In the 1970s, noted economist John Kenneth Galbraith provided some measures of concentrated power:

In 1976, the five largest industrial corporations, with combined assets of $113 billion, had just under 13 percent of all assets used in manufacturing. The 50 largest manufacturing corporations had 42 percent of all assets. The 500 largest had 72 percent.

In the same year corporations with assets of more than a billion dollars, 162 in all, had 54 percent of all assets in manufacturing; corporations with assets of more than $100 million had approximately four-fifths of assets; and 3801 firms with assets of more than $10 million had 89 percent of all assets.[28]

Such concentrated power means that the production, marketing and other decisions made in a handful of enterprises can rapidly change entire industries and affect the entire world.

———————✣———————

In the first four decades of the twentieth century, the *fossil economy*, which Andreas Malm defines as "an economy of self-sustaining growth predicated on the growing consumption of fossil fuels, and therefore generating a sustained growth in emissions of carbon dioxide,"[29] became thoroughly entrenched in all of the advanced capitalist countries, and established strong footholds in what would later be called the Third World.

And yet, even a quick look at the Great Acceleration graphs that directly reflect fossil fuel use—carbon dioxide, nitrous oxide, methane, real GDP, primary energy use, and transportation—shows that fossil fuel had barely begun to achieve its potential before World War II. In Barry Commoner's words, "We know that *something* went wrong in the country after World War II, for most of our serious pollution problems either began in the postwar years or have greatly worsened since then."[30]

9

War, Class Struggle, and Cheap Oil

The tens of millions of dead in the two World Wars brought about tens of trillions of profitable investments in the huge reconstructions of destroyed homes and industries and ongoing rearmament: a million dollars or more per dead body.

—DARKO SUVIN[1]

The habit of using historical events as markers can distract us from the actual content of those events. When we say that the Anthropocene began after World War II, we usually mean only that it started after 1945—but what we ought to mean is that it followed the most destructive, most murderous, most inhumane conflict in all history.

That's an important distinction because the war wasn't just a passive divider between Holocene and Anthropocene. World War II and its aftermath created the conditions that have shaped capitalism ever since, and started the Great Acceleration on its environmentally destructive course. This chapter considers the war and its consequences as a time of transition between two epochs.

The Profits of War

The development of capitalism since the early 1800s was marked by increasing dependence on fossil fuels, but the pace of economic growth and change was slowed by the Great Depression of the 1930s, and further interrupted by World War II. Given capitalism's inherent need not just to grow, but to grow faster, it's possible that something like the Great Acceleration would have occurred even if the Depression and war didn't happen, but that is just speculation. In the real world, capitalism's worst depression and most destructive war set the stage for the economic and social changes that have pushed the Earth System into a new and dangerous epoch.

It is impossible to overstate the horror of the Second World War, a conflict in which competing empires used every possible resource and weapon in a fight for global dominance. In six years, 60 million soldiers and civilians were killed by military action or state-organized genocide, and another 20 million died of hunger and disease. Twenty-seven million people died in the USSR. Three million starved to death in India, victims of what's been called "Winston Churchill's secret war." Entire cities in Japan and Europe were leveled. Vast forests, millions of hectares of farmland, and billions of dollars' worth of industrial plants were destroyed. In August 1945, the most destructive weapon ever made killed over 100,000 people in two brief moments, and condemned uncountable others to lingering deaths.

When the fighting stopped, huge areas of Europe, Asia, and Africa were in ruins, their economic and physical infrastructures leveled. Britain was in better shape than most of Europe, but its economy was weak, and its government was nearly broke.

The only winner was the United States. It emerged from the war physically unscathed and economically more powerful than ever. War production had nearly doubled the country's GNP: almost two-thirds of the world's industrial production was concentrated in that one country, the new global hegemon.[2]

Gains for Monopoly Capital

In 1942, Roosevelt's secretary of war Henry Stimson, explained why he opposed an excess profits tax: "If you are going to prepare for war in a capitalist country, you have to let business make money out of the process, or business won't work."[3] So while millions were slaughtered, a few got very rich. As a socialist writer commented in 1946: "For the American plutocracy, the Second World War was the most profitable enterprise in its whole career. It made the American capitalists the richest rulers that had ever emerged in human history."[4]

The long-term trends that we discussed in chapter 8 were greatly strengthened during the war.

Corporate Concentration. Of the $175 billion in war production contracts awarded during the war, two-thirds went to a hundred companies, and more than half to just 33. Nearly 80 percent of the new factories built with government money were operated by the 250 largest corporations. After the war, those facilities were sold for less than a quarter of what they had cost to build, and 87 companies acquired two-thirds of them. Most contracts were cost-plus, so corporate profits were guaranteed. During the war, U.S. corporations made $52 billion in after-tax profits, accumulated some $85 billion in capital reserves, and added more than 50 percent to their productive capacity.[5]

By the end of the war, 31 percent of U.S. workers were employed in corporations with over 10,000 employees, compared to just 13 percent in 1939.[6] In the same period, companies with under 500 employees fell from 52 percent to 34 percent. In 1946, the Senate Small Business Committee reported that the 250 largest corporations controlled "66.5 percent of total usable facilities and almost as much as the entire 39.6 billion dollars held before the war by all the more than 75,000 manufacturing corporations in existence."[7]

U.S. corporations began the postwar period with a huge reservoir of cash, and productive infrastructure that was both newer and larger than anything available to potential competitors in other countries.

Oil and Automobiles. Oil had been important in World War I; in World War II, it was decisive:

> Far more than they had in any previous conflict, oil-powered weapons—tanks, airplanes, submarines, aircraft carriers, and armored troop carriers—dominated the theaters of war. . . . The demands for fuel were prodigious: a typical armored battalion required seventeen thousand gallons of oil to travel only one hundred miles; the U.S. Fifth Fleet alone consumed 630 million gallons of fuel oil during a single two-month period.[8]

Six out of every seven barrels of oil used by the Allied forces came from U.S. wells and were refined by U.S. oil companies.[9] To ensure supply, the government built pipelines to carry oil from Texas to refineries in the Northeast, and, in "one of the largest and most complex industrial undertakings of the war," built dozens of new refineries equipped with new technology that produced 100-octane fuel for airplanes.[10] After the war, high-octane gasoline would power both energy-intensive production technology and the V8 engines in Detroit's absurdly large passenger cars.

Would there be enough oil left for the postwar economy, let alone for another war? No one knew how much was still in the ground in the United States, so in 1943 the government took steps to avoid future shortages by bribing the absolute feudal monarch Ibn Saud to give exclusive rights to Saudi oil to a consortium of U.S. oil corporations.

U.S. production of passenger cars, commercial trucks, and auto parts stopped entirely in 1942, but U.S. automakers prospered, receiving some $29 billion to produce over three million jeeps and trucks, as well as airplane engines, tanks, armored cars, machine guns, and bombs. They ended the war with their facilities not just intact but updated and expanded.

Industrial Chemistry. During the Second World War, the U.S. government spent over $3 billion to build or expand petrochemical plants to produce nitrogen for explosives, synthetic rubber for tires, nylon for parachutes, and more. After the war, oil and chemical companies bought those factories at bargain basement prices:

among others, a $2 million plant was sold for $325,000 (Standard Oil); a $19 million plant for $10 million (Monsanto); and a $38 million operation for $13 million (DuPont).[11]

That giveaway was the basis of the Age of Plastics, a time when DuPont could use the slogan "Better Things for Better Living Through Chemistry" without being ridiculed. As historian Kevin Phillips writes, U.S. manufacturing technology "had been revolutionized by wartime demands" with the result that "executives in company after company found themselves selling products commercially unfeasible before Pearl Harbor."[12] From near nonexistence before the war, plastics grew to become the third-largest manufacturing industry in the United States, a status it still claims today.

Military Keynesianism

In February 1944, the magazine *Politics* featured an article by socialist Walter J. Oakes, predicting that after the war, in contrast to previous wars, the United States would maintain a high level of military spending. The ruling class, he said, had two major postwar objectives: to begin preparations for World War III, and to prevent the social unrest that would occur if massive unemployment returned. To achieve those goals, the United States would enter "the epoch of Permanent War Economy."[13]

It took longer to implement than Oakes predicted, but his forecast was broadly correct. In 1950, *U.S. News and World Report* informed its business readers that "government planners figure that they have found the magic formula for almost endless good times. . . . Cold War is an automatic pump primer."[14] Military Keynesianism—massive military spending to maintain or increase economic growth—has been a fundamental feature of the U.S. economy, no matter which political party was in office, for over half a century.

In the 1930s, the U.S. military budget was about $500 million a year. Despite cutbacks when the war ended, from 1946 through 1949 spending on military personnel and weapons was 38 times the prewar level, averaging more than $19 billion a year.[15] It jumped substantially when the United States intervened in the Korean civil war, a "police action" that killed more than 2 million people, most of them civilians:

> Truman's military buildup in the opening months of the Korean conflict exceeded even the mobilization in the early days of World War II. He increased the number of troops from 1.5 million to 3.2 million, of army divisions from 10 to 18, of air force wings from 42 to 72, and the number of ships from 618 to 1,000, including fourteen carrier groups— in the first year of the Korean War. Congress allocated $50 billion to do the job, and Truman sought $62.2 billion for the next year—and still had to ward off the Joint Chiefs of Staff, who wanted more than $100 billion. Only about 25 percent of these huge sums was meant for the Korean War—most of the money was earmarked for the global struggle against communism.[16]

The hundreds of billions of dollars that were pumped into arms production and related industries during the Korean War, and in every year since, meant more capital investment in high-energy, high-polluting factories, and, indirectly, more consumer spending on cars, homes, appliances and more. Those billions also had, as we'll see in chapter 10, horrendous environmental impacts.

Reconversion and Class Struggle

Liberal historians like to present the postwar years as a rapid, virtually seamless transition from wartime austerity to the long boom. Thanks to the wise policies of the Truman government, and the accumulated savings of workers and soldiers, the "reconversion" brought universal prosperity without disruption. In his pioneering history of the CIO, Marxist historian Art Preis disputed that:

This falsification is accomplished by lumping the years of the Korean War boom with the preceding years of stagnation and decline. It hides the real conditions that prevailed during the peacetime years of Truman's administration. It tends to cover up the most vital fact of modern American economic history: At no time since 1929 has American capitalism maintained even a semblance of economic stability and growth without huge military spending and war debt.[17]

Political economist Lynn Turgeon agrees:

Although the postwar economy of the United States was stabilized by the pent-up demand and forced savings built up during the war, the Marshall Plan and Point 4 and its successor, Foreign Aid (later the Agency for International Development), the transition to a peacetime economy was remarkably sluggish. Real income in 1950 was little higher than it had been in 1945, and it wasn't until the Korean War boom that the economy overtook the annual production at the end of World War II.[18]

Pent-up consumer demand may have helped to prevent recession in a statistical sense, but the bosses' prosperity didn't trickle down to most working people until the 1950s. For many, the years immediately after the war were particularly hard. After Germany surrendered in May 1945, U.S. manufacturers started cutting back working hours and laying off workers, and the removal of wartime controls inflated prices of food and other essentials. By October 1945, nearly two million workers were unemployed, and real incomes had fallen 15 percent.[19] Women were particularly hard hit as employers insisted on returning to a normal—meaning all-male—workforce.

Those developments, and four years of accumulated grievances, triggered the biggest strike wave in U.S. history. In 1945, 3.5 million workers took part in 4,750 strikes, and in 1946, 4.6 million workers took part in 4,985 strikes. According to Art Preis, "For the number of strikers, their weight in industry and the duration of the struggle, the 1945–46 strike wave in the U. S. surpassed anything of its kind in any capitalist country, including the British General Strike of 1926."[20]

As noted business consultant Peter Drucker wrote in 1946, the national leaders of the big unions were not happy about the strike wave: "In the five major strikes of the first postwar winter, 1945–46—the General Motors strike, the meat packers' strike, the steel strike, the electrical workers' strike and the railway strike—it was on the whole not the leadership which forced the workers into a strike but worker pressure that forced a strike upon a reluctant leadership."[21]

Class struggle from below met class struggle from above. The response of the bosses and their government involved a mixture of carrots and sticks.

- They passed the Taft-Hartley Act to outlaw wildcat strikes and most other "irresponsible" labor actions, banned closed shops, and put major legal barriers in the way of organizing new groups of workers.
- They helped conservative and liberal union leaders consolidate their authority through a destructive red scare campaign that split the union movement and suppressed militants.
- President Truman repeatedly broke strikes by using wartime legislation to seize control of mines, refineries, and other industries.
- Uncooperative labor leaders and unions were punished: most notably, the United Mine Workers was fined $3.5 million, and leader John L. Lewis was personally fined $10,000 for refusing to call off a coal strike.
- After the strike wave ebbed, auto companies and other major manufacturers negotiated pay increases and job security in exchange for multiyear contracts that guaranteed uninterrupted production.

By 1950, the U.S. ruling class had thoroughly weakened its only potential domestic opposition: instead of militant workers, American capitalism could now count on "an ideologically loyal and industrially moderate labor movement, which bargained for concessions without challenging basic dispositions of a business society, [which] was more or less accepted as a constructive addition to the team in a period of cold war confrontations and economic expansion."[22] Michael Yates sums up this essential precondition of the long boom:

> Corporate and public leaders engineered a peace accord with "legitimate" labor leaders. If labor unions would cede to employers the right to run the plants, to set prices, and unilaterally introduce machinery, and if union leaders would enforce no-strike agreements and discipline unruly members, employers would guarantee steady increases in wages and the introduction and growth of fringe benefits, and they would accept the unions as legal representatives of employees. They would also not block the passage of social welfare legislation.[23]

Converting Europe to Oil

For two years after the war ended, U.S. policy toward Germany and Japan was punitive: it aimed to destroy any possibility that either could become a major economic power again. Many of those countries' surviving factories were physically dismantled and shipped away, while others faced severe restrictions on the volume and nature of goods they could produce.

It soon became clear, however, that such policies were inconsistent with U.S. foreign policy, which aimed to build an alliance of capitalist states to "contain communism." In 1948, Congress approved the European Recovery Program, commonly called the Marshall Plan. Over the next three years it gave $13 billion to European governments, including West Germany. That's equal to about $130 billion today, and far more as a share of U.S. GDP. It wasn't just the biggest U.S. foreign aid plan ever, it was bigger than all previous foreign aid plans combined.[24]

Many writers have portrayed the Marshall Plan as an example of great-power benevolence, in which the United States selflessly assumed responsibility for restoring prosperity in Europe, but it was nothing of the kind. The Marshall Plan's purpose was to strengthen U.S. corporations—especially oil companies—in the United States and internationally. Most Marshall Plan money had to be used for purchases from U.S. corporations—while its importance in European reconstruction shouldn't be overlooked, in many ways it served to boost the U.S. economy with indirect government purchases from U.S. corporations.

Within a few months of the plan's launch, an exposé in the *Chicago Tribune* showed what that often meant in practice:

> The *Tribune* has examined official records of Marshall Plan business for the 45-day period ended September 15—one-eighth of the first year of the plan. It appears, for example, that the AngloAmerican Oil Company, Ltd. obtained permission from the British government to go shopping for petroleum products in America. The record shows that it was able to buy $7,258,332 worth of products from the Esso Export corporation and the Standard Oil Export corporation, both of New York. Thus the British concern got the oil it was after, and was doubtless able to make a good profit selling it to its customers. The two American companies were paid in dollars in New York for the oil supplied to the British companies, likewise on a remunerative basis.
>
> What makes the transactions notable is that the British buyer, AngloAmerican, is owned 100 percent by the Standard Oil Company of New Jersey. The American sellers, Esso Export and Standard Oil Export, are also owned 100 percent by Standard of New Jersey. Thus what the Marshall Plan actually did was to enable the biggest American oil company to

shift some merchandise from one department to another, collecting two profits on the operation, at the expense of the American taxpayer.[25]

The *Tribune* noted that representatives of the Rockefeller family, principal owners of Standard Oil Company of New Jersey (later renamed Exxon), were among the strongest forces that lobbied Congress to approve the Marshall Plan. As it developed, the Rockefellers and other oil billionaires were among the plan's prime beneficiaries.

Between 1948 and 1951, more than half of the oil sold to Western European buyers by U.S. oil companies was paid for with Marshall Plan funds.[26] Oil accounted for 10 percent of all Marshall Plan spending—20 percent in 1949—far more than was allocated to any other commodity. Even an historian who is very favorable to the plan laments that "Marshall Planners treated the interests of the country and the oil industry interchangeably." European buyers were charged above-market prices for oil, and they had to buy gasoline and diesel rather than less expensive crude oil, because the Marshall Plan refused "to finance projects that threatened to compete with U.S. companies," such as rebuilding European refineries that were damaged in the war.[27]

The U.S. Congress had specified that U.S. oil should not be used under the Marshall Plan. This restriction was justified as protection for domestic consumption, but it had the effect of subsidizing the expansion of U.S. oil companies' then-new facilities in Saudi Arabia, and reshaping Europe's energy use patterns. Before the war, only 20 percent of European oil imports had come from the Middle East, but that rose to 43 percent in 1947 and 85 percent by 1950, and that accelerated Europe's long-term transition from dependence on coal to dependence on oil. "In 1955, coal provided 75 percent of total energy use in Western Europe, and petroleum just 23 percent. By 1972, coal's share had shrunk to 22 percent, while oil's had risen to 60 percent—almost a complete flip-flop."[28]

Before World War II, U.S. federal and state governments had promoted domestic oil production directly through depletion allowances that minimized their taxes, and indirectly by immense road-building projects throughout the country. In the 1940s, the geography of oil support policies expanded: the Marshall Plan's subsidies to U.S. oil operations in the Middle East initiated a permanent policy of treating Middle East oil as central to U.S. foreign policy.

Cheap and Abundant Oil

Although no one yet knew how big it was, production was about to start in what turned out to be the largest conventional petroleum deposit in the world. In the next six decades, about 60 billion barrels of crude oil would be extracted in Ghawar, Saudi Arabia, and it wasn't just abundant, it was cheap. "One of the reasons why the Golden Age was golden was that the price of a barrel of Saudi oil averaged less than $2 throughout the entire period from 1950 to 1973, thus making energy ridiculously cheap, and getting cheaper all the time."[29]

The Great Acceleration would not have been possible without cheap oil—as a commodity in its own right, as the raw material for plastics and other petrochemicals, as the enabler for high-energy manufacturing processes, and above all as the fuel for hundreds of millions of cars, trucks, ships, and planes.

Total world energy consumption more than tripled between 1949 and 1972. Yet that growth paled beside the rise in oil demand, which in the same years increased more than five and a half times over. Everywhere, growth in the demand for oil was strong. Between 1948 and 1972, consumption tripled in the United States, from 5.8 to 16.4 million barrels per day—unprecedented except when measured against what was happening elsewhere. In the same years, demand for oil in

Western Europe increased fifteen times over, from 970,000 to 14.1 million barrels per day. In Japan, the change was nothing less than spectacular; consumption increased 137 times over, from 32,000 to 4.4 million barrels per day.[30]

Between the Second World War and 1973, the fossil economy was solidified and globalized in the Global North:

> Between 1946 and 1973 the world consumed more commercial energy than had been used in the entire period from 1800 to 1945. While the world consumed around 53 billion tons of oil equivalent of energy in the 1800–1945 period, over 84 billion tons of oil equivalent were used in the twenty-seven years that followed the war. . . .
>
> These years witnessed the emergence of large-scale natural gas and nuclear power industries, as well as a recovery of world coal production. The most dynamic energy industry of all, however, was oil. Indeed, world oil production grew by more than 700 percent in the period 1946–73.[31]

Anticipations from the Left

The idea that a major change in the relationship between human society and the global environment occurred after the Second World War is not new to the left. Although none of the many scientific papers and talks about the Great Acceleration mention it, some of their conclusions were anticipated decades ago, by three founders of radical environmentalism.

Rachel Carson, in *Silent Spring* (1962):

> For the first time in the history of the world, every human being is now subjected to contact with dangerous chemicals, from the moment of conception until death. In the less than

two decades of their use, the synthetic pesticides have been so thoroughly distributed throughout the animate and inanimate world that they occur virtually everywhere. . . .

All this has come about because of the sudden rise and prodigious growth of an industry for the production of man-made or synthetic chemicals with insecticidal properties. This industry is a child of the Second World War.[32]

Murray Bookchin, in *Our Synthetic Environment* (1962):

Since World War II . . . there has been a new industrial revolution, and the problems of urban life have acquired new dimensions. . . . At the same time that the number of pollutants has increased, the ecological preconditions for wholesome air and plentiful water are being undermined.[33]

Barry Commoner, in *The Closing Circle* (1971):

The chief reason for the environmental crisis that has engulfed the United States in recent years is the sweeping transformation of productive technology since World War II. . . . Productive technologies with intense impacts on the environment have displaced less destructive ones. The environmental crisis is the inevitable result of this counter-ecological pattern of growth.[34]

In 1991, Commoner anticipated Crutzen's argument about the growth of human impact on the Earth System: "The technosphere has become sufficiently large and intense to alter the natural processes that govern the ecosphere."[35]

And in 1994, ten years before the IGBP's pathbreaking synthesis was published, John Bellamy Foster updated Commoner's argument, and initiated a Marxist analysis of the social and economic changes that caused what would later be dubbed the Great Acceleration:

In the period after 1945 the world entered a new stage of planetary crisis in which human economic activities began to affect in entirely new ways the basic conditions of life on earth. This new ecological stage was connected to the rise, earlier in the century, of monopoly capitalism, an economy dominated by large firms, and to the accompanying transformations in the relation between science and industry. Synthetic products that were not biodegradable—that could not be broken down by natural cycles—became basic elements of industrial output. Moreover, as the world economy continued to grow, the scale of human economic processes began to rival the ecological cycles of the planet, opening up as never before the possibility of planet-wide ecological disaster. Today few can doubt that the system has crossed critical thresholds of ecological sustainability, raising questions about the vulnerability of the entire planet.

"What transpired in the post–World War II period," Foster wrote, was "a qualitative transformation in the level of human destructiveness."[36]

10

Accelerating into the Anthropocene

We are coming to a fork in the road in human history, where the system of global capitalism is forcing an end to the Holocene Epoch of the last 12,000 years, the geological period within which human civilisation has developed, where we have to decide between 'capitalism or the planet'.

—DEL WESTON[1]

At the beginning of 1950, four key drivers of the long boom were in place: a powerful industrial base in the United States, concentrated in a few hundred giant corporations and dominated by the petroleum/automotive sector; a large and growing military budget; a disciplined and financially secure labor force, purged of militants and militancy; and a seemingly infinite supply of cheap energy. The atmospheric concentration of carbon dioxide had risen about 35 parts per million above the preindustrial level, and 65 percent of that increase was due to emissions generated in just two countries, the United States and the United Kingdom.[2]

From an Earth System perspective, history since then can be told as an account of the expansion of fossil capitalism into every aspect of life and every part of the globe. The Great Acceleration graphs produced by the IGBP show the effects of that expansion taking off about 1950, and speeding up in virtually every respect ever since.

This chapter begins with an overview of the so-called Golden Age, and then considers key trends that shaped economic and environmental change in the second half of the twentieth century and drove the Anthropocene transition.

A Semi-Golden Age

The period from after the war to 1973 is frequently called the Golden Age of Capitalism, or just the Golden Years. Economic slowdowns did occur in those years, but they were brief—overall, it was the longest continuous economic boom in capitalist history. It is frequently described in glowing terms, even by authors with unquestioned anti-capitalist credentials. A case in point:

> The lives of vast numbers of people were transformed. Unemployment fell to levels only known before in brief periods of boom—3 percent in the U.S. in the early 1950s, 1.5 percent in Britain, and 1 percent in West Germany by 1960. There was a gradual and more or less uninterrupted rise in real wages in the U.S., Britain, and Scandinavia in the 1950s, and in France and Italy in the 1960s. Workers were living better than their parents, and expected their children to live better still.
>
> It was not just a question of higher incomes. Wages could be spent on a range of consumer goods—vacuum cleaners, washing machines, refrigerators, televisions, instant hot water systems. There was a qualitative leap in the working-class standard of living. Housework remained a chore for women, but no longer entailed endless hours of boiling and kneeling and scrubbing. Food could be purchased weekly rather than daily (opening the door for the supermarket to replace the corner shop). Entertainment of sorts was on tap in the home, even for those who could not afford the cinema, theatre or dancehall.[3]

Such accounts are broadly accurate, but must be qualified, because only a privileged minority experienced such prosperity, even in the richest country on earth. Mike Davis estimates that about a quarter of the U.S. working class, mainly "white-ethnic semi-skilled workers and their families," were able to join the middle-class home-in-the-suburbs life portrayed in *Father Knows Best* and other popular television programs. "Another quarter to one-third of the population, however, including most blacks and all agricultural laborers, remained outside the boom, constituting the 'other America' which rebelled in the 1960s."[4]

Across the United States, this period saw the formation of the deep *geographic divide* that left-wing journalist Michael Harrington described in his 1962 bestseller, *The Other America*:

> Now the American city has been transformed. The poor still inhabit the miserable housing in the central area, but they are increasingly isolated from contact with, or sight of, anybody else. Middle-class women coming in from Suburbia on a rare trip may catch the merest glimpse of the other America on the way to an evening at the theater, but their children are segregated in suburban schools. The business or professional man may drive along the fringes of slums in a car or bus, but it is not an important experience to him. The failures, the unskilled, the disabled, the aged, and the minorities are right there, across the tracks, where they have always been. But hardly anyone else is.
>
> In short, the very development of the American city has removed poverty from the living, emotional experience of millions upon millions of middle-class Americans. Living out in the suburbs, it is easy to assume that ours is, indeed, an affluent society.[5]

This racially and economically enforced segregation laid the basis for the inner-city rebellions that swept the United States in the mid- and late-1960s, shattering the myth of universal prosperity.

The "golden" label is even less appropriate on a global scale. Historian Eric Hobsbawm reminds us that "the Golden Age essentially belonged to the developed capitalist countries," home to 75 percent of global production and over 80 percent of manufactured exports. "General affluence never came within sight of the majority of the world's population."[6] That was and remains true of the Great Acceleration: as the 2015 update showed, it has been "almost entirely driven by a small fraction of the human population, those in developed countries."[7]

Nevertheless, the emergence of a large, relatively privileged segment of the working class in the United States played a big role in keeping the long boom going, and in making petroleum use a fundamental feature of life in the advanced capitalist countries.

Cars and Suburbs

The top two items on many of those workers' shopping lists, after years of shortages and privation, were cars and houses. In its 1950 *Annual Report*, the Continental Oil Company (Conoco) told its shareholders that "the major contributing factor to the high level of business activity of the country was the continued high demand for housing and automobiles, two industries which vitally affect the consumption of petroleum products."[8]

Most U.S. industries didn't convert to civilian production until 1946 or 1947, but gasoline rationing ended on August 15, 1945— one day after Japan surrendered—and the first civilian car built since 1942 rolled off the Ford assembly line just two months later. The auto-oil complex was counting on pent-up demand and the impact of immense spending on advertising, and the bet paid off. In 1945 there were 26 million cars on the road in the United States; five years later there were 40 million.[9] The second wave of automobilization was even bigger, and had a greater effect on the economy as a whole than the first:

> From 1947 through 1960 the motor vehicle, petroleum, and rubber industries were responsible for one-third of all plant and equipment expenditures in manufacturing. Consumer outlays on automobiles and parts, gasoline, and oil rose from

6.5 percent of total expenditures to 9 percent during the same period. By 1963–66 one of every six business enterprises was directly dependent on the manufacture, distribution, servicing, and use of motor vehicles; at least 13.5 million people, or 19 percent of total employment, worked in "highway transport industries." Automobile registrations climbed from 25.8 million in 1945 to 61.7 million in 1960. . . . In the late 1980s the effects were still evident, as nine of the top 14 Fortune 500 largest industrial corporations in 1987, ranked by sales, were automobile or oil companies.[10]

The automobile boom was both cause and effect of urban sprawl, more politely dubbed suburbanization. If you owned a car, living in the suburbs was possible; if you bought a home in the suburbs, owning a car was essential.

After the war, contractors who had learned the techniques of mass housing construction on military bases moved quickly to get their share of workers' savings and veterans' government guaranteed housing loans. Annual housing starts increased from 142,000 in 1944 to over a million in 1946 to nearly 2 million in 1950, stabilizing at 1.3 million a year after that. One out of every four homes standing in 1960 had been built in the 1950s.[11] Over 80 percent of the new houses were single-family homes built outside the cities, where land was cheap, building regulations were less strictly enforced, and mass production of cookie-cutter houses was easier to manage.

Public transportation in the new suburbs was poor or non-existent, and instead of neighborhood shops there were distant shopping centers, so house-ownership and car-ownership went together. And this was only part of what David Harvey calls the "wants, needs, and desires associated with the rise of a suburban lifestyle":

Not only are we talking about the need for cars, gasoline, highways, suburban tract houses, and shopping malls, but also lawn mowers, refrigerators, air-conditioners, drapes, furniture (interior and exterior), interior entertainment equipment (the TV), and a whole mass of maintenance systems to keep this daily life going. Daily living in the suburbs required the consumption of at least all of that. The development of suburbia turned these commodities from wants and desires into absolute needs.[12]

The automobile boom was also cause and effect of a long-term boom in road-building, including the $114 billion (over $400 billion in 2016 dollars) Interstate Highway System, which was justified as essential for moving troops and military equipment if the United States were ever invaded. In 2001, Lester Brown calculated that the country had 3.9 million miles of roads, "enough to circle the earth at the equator 157 times. . . . The U.S. area devoted to roads and parking lots covers an estimated 61,000 square miles, an expanse approaching the size of the 51.9 million acres that U.S. farmers planted in wheat last year."[13]

The automobile/suburbia boom that began in the 1940s headed off a new depression and provided jobs and homes for millions, but those benefits had a deadly cost. From an environmental perspective the automobile-suburbia boom may well be, as author and social critic James Kunstler believes, "the greatest misallocation of resources in the history of the world":

There really is no way to fully calculate the cost of doing what we did in America, even if you try to tote up only the monetary costs (leaving out the social and environmental ones). Certainly it is somewhere up in the tens of trillions of dollars when one figures in all the roads and highways, all the cars and trucks built since 1905, the far-flung networks of electricity, telephone, and water lines, the scores of thousands of housing subdivisions, a similar number of strip malls, thousands of regional shopping malls, power centers, big-box pods, hamburger and pizza shacks, donut shops, office parks, central schools, and all the other constructed accessories of that life. . . .

More than half the U.S. population lives in it. The economy of recent decades is based largely on the building and servicing of it. And the whole system will not operate without liberal and reliable supplies of cheap oil and natural gas.[14]

Industrial Agriculture

"Farming is growing peanuts. Agriculture is turning petroleum into peanut butter." In those few words, biologist Richard Lewontin summed up the transformation of food production in the Great Acceleration. Just as concisely, his colleague Richard Levins explained why the change took place: "Agriculture is not about producing food but about profit. Food is a side effect."[15]

The integration of food production into the fossil economy began before the Second World War in North America, when gasoline- or diesel-powered tractors began to replace animal power. Farmers who could afford the new machines could farm larger areas with fewer farmworkers: they could produce more and sell for less. Those who could not were squeezed out when prices fell below their costs. Between 1930 and 1945, the number of working animals (horses and mules) on U.S. farms fell from 18.7 million to 2.4 million; the number of tractors increased from 920,000 to 2.4 million; and the agricultural workforce fell from 21.5 to 16 percent of all workers.[16] And that was only the beginning.

By 1960 there were 4.7 million tractors and only 3 million work animals on farms, and after that there were so few work animals that the U.S. Census Bureau stopped counting them. There are still about 2 million family-owned farms in the United States, but most survive only because the owners have off-farm jobs. Just 11 percent of farms now account for 85 percent of farm output.[17] Agricultural ecologists Ivette Perfecto, John Vandermeer, and Angus Wright describe the transformation:

Starting with the early mechanization of agriculture that sub-
stituted traction power for animal power, to the substitution
of synthetic fertilizer for compost and manure, to the substitution of pesticides for cultural and biological control, the
history of agricultural technological development has been a
process of capitalization that has resulted in the reduction of
the value added within the farm itself. In today's farms, the
labour comes from Caterpillar or John Deere, the energy from
Exxon/Mobil, the fertilizer from DuPont, and the pest management from Dow or Monsanto. Seeds, literally the germ that
makes agriculture possible, have been patented and need to be
bought.[18]

This transformation was the equivalent, in farming, of the
transformation that took place in manufacturing in the 1800s: a
shift from petty commodity production to what Marx called "a
specifically capitalist mode of production," along with concentrated ownership, and dependence on fossil fuels. In his account
of the role of petroleum in American life, Matthew Huber writes:

Since World War II, the American food system has been completely *fossilized*, or has grown to rely on fossil fuel inputs at
each stage of production, distribution, packaging, and consumption. This is especially true of agricultural production
where mainly biological forms of energy (muscles) were
replaced by inanimate fossil machine power and energy-intensive synthesized chemical inputs (fertilizers and pesticides) in
place of natural ones (manure).[19]

It now takes more energy to produce food than we obtain from
eating it: every calorie of food energy requires 10 calories of fossil
energy. In 2007, the U.S. food system, from farm to table, used 16
percent of all of the country's energy.[20] The Haber-Bosch process,
discussed in chapter 4, is a particular energy hog: not only is more

energy used to extract nitrogen from the air than in manufacturing farm machines, but it takes 33,500 cubic feet of natural gas
(methane) to produce a ton of nitrogen-based fertilizer.[21] It's been
estimated that 1 percent of total global energy is devoted to producing synthetic nitrogen fertilizers.[22]

Industrial agriculture also requires high levels of pesticides,
largely derived from petroleum. In 2007, some 5.2 billion pounds
of pesticides were applied worldwide. Studies have shown that only
0.1 percent of pesticides used in agriculture actually reach their
targets: the rest end up in soil, air, and groundwater. Increasingly,
the target pests are resistant to pesticides, leading to a "pesticide
treadmill" in which ever larger applications of increasingly toxic
chemicals are needed to produce the same results.[23]

Military Pollution

Officially, the U.S. military budget in 2015 was $598.5 billion, but
the official figures, as Foster, Holleman, and McChesney have
shown, exclude such clearly military items as veterans' benefits,
military grants to foreign governments, and interest payments on
military debts. When those are added in, the total rises to over one
trillion dollars a year.[24] That's appalling in its own right; it's worse
when we realize that it is more than the combined military budgets
of the next nine biggest military spenders in the world. Analysts at
the Trans National Institute calculate that global military spending
in 2013 totaled $1.7 trillion—"130 times that of planned humani-
tarian spending and dwarfing any investment in climate change."[25]

As we've seen, military spending provided a big boost to economic growth in the 1940s and 1950s, and has helped to maintain
growth since then. It has also created a specific and ever-growing military market for fossil fuels, especially oil. Today the U.S.
military is the world's largest user of petroleum, *and* the largest
polluter, producing more hazardous waste than the five largest
U.S. chemical companies combined, *and* the largest producer of
greenhouse gases.[26]

Vaclav Smil estimates that during the 1990s the U.S. military consumed "more than the total commercial energy consumption of nearly two-thirds of the world's countries"—and that excludes fuel used in the 1991 Gulf War and the 1998 bombings of Serbia and Kosovo.[27] It's impossible to determine the impact of the U.S. military on climate change, because military emissions were excluded from the Kyoto Accord.[28] Researcher Barry Sanders, who has done more than anyone to dig out the facts, says the official figure of 395,000 barrels a day omits a great deal: according to his estimates, one million barrels a day is "a safe, and even conservative number." That's enough to increase total U.S. emissions by an outrageous 5 percent.[29]

As investigative journalist Sonia Shah says, the U.S. military seems to make no effort to reduce its use of fuel:

> The Army employed sixty thousand soldiers solely for the purpose of providing petroleum, oil, and lubricants to its war machines, which have themselves become increasingly fuel-heavy. The sixty-eight-ton Abrams tank, for instance, burns through a gallon of fuel for every half mile. With its inefficient, 1960s-era engine, the Abrams tank burns twelve gallons of fuel an hour *just idling.*
>
> So much time and money is spent fueling the American fighting machines that, according to the head of the Army Materiel Command, a gallon of fuel delivered to the U.S. military in action can ultimately cost up to $400 a gallon. Indeed, 70 percent of the weight of all the soldiers, vehicles, and weapons of the entire U.S. Army is pure fuel.

In 2001, a Defense Science Board panel concluded that continuing to support the military's demand for fuel would require either more oil-efficient weapons systems or bigger support systems. Shah comments that the generals seem to have chosen a third option: capturing access to more oil.[30]

Of course, greenhouse gas emissions are only part, and possibly not the largest part, of the damage caused by military action. For most industries, environmental destruction is a side effect of the endless quest for profit, but the military and the industries that serve it profit directly from destruction. The arms industry uses capitalist means of production to create ever greater forces of destruction.

This led political economist Peter Custers to propose that the concept of *negative use-value* be incorporated in Marxist economic theory, to account for "the adverse health and environmental consequences of capitalist production, and, most particularly, the damaging properties of military and nuclear commodities."[31] Similarly, in 1860, the English philosopher and art critic John Ruskin coined the word *illth* for accumulation that causes harm, as opposed to *wealth*, which he thought should only be used for things that promote social well-being. Military products and actions certainly deserve to be called illth.

Patricia Hynes, a retired professor of environmental health and chair of the Traprock Center for Peace and Justice, documented the environmental consequences of U.S. militarism in *Pentagon Pollution*, a series of articles that *Climate & Capitalism* published in 2015. She writes:

> Modern war and militarism have a staggering impact on nature and our lived environment—by the kinds of weapons used (long-lived concealed explosives, toxic chemicals, and radiation); the "shock and awe" intensity of industrial warfare; and the massive exploitation of natural resources and fossil fuels to support militarism. By 1990, researchers estimated that the world's military accounted for 5–10 percent of global air pollution, including carbon dioxide, ozone-depletion, smog, and acid-forming chemicals. The Research Institute for Peace Policy in Starnberg, Germany, calculated that 20 percent of all global environmental degradation was due to military and related activities.[32]

Any serious effort to stop global warming will have to overcome the resistance of the U.S. military. As Barry Sanders writes:

"Today's brand of warfare, especially, just relies too heavily on staggering amounts of oil; it produces too much greenhouse gas to ever willingly constrain itself with any kind of treaty."[33]

Globalized Production

The length of the long boom led many mainstream economists to conclude that the business cycle had been overcome, but capitalism's essential dynamics always assert themselves, and every golden age eventually turns to lead. In January 1973, New York Stock Market prices started falling: they eventually dropped 45 percent. In October 1973, OPEC imposed increased taxes on oil companies, triggering an almost-immediate fourfold increase in the price of oil. Partly as a result of that, and partly as an extension of a general profit decline that had been under way since 1970, the economy went into freefall. The 1974–75 economic crisis wasn't just another slowdown: it was the first recession since the 1930s to strike all the imperialist countries simultaneously, "a full-fledged structural crisis, ending the long boom, and marking the beginning of decades of deepening stagnation."[34] In the same period, the United States lost the war in Vietnam, a massive revolutionary upsurge overthrew the dictatorship that had ruled Portugal for thirty-six years, and in six years insurrectionary movements seized power in fourteen Third World countries, including Nicaragua, Mozambique, Angola, and Iran.

The more farsighted sections of the ruling classes in most countries saw a need to change direction: the policies that had maintained capitalist growth since the Second World War were no longer working. The path the largest corporations adopted is usually called "neoliberalism," a word of somewhat fuzzy meaning. For our discussion, three elements are significant: all involved corporate efforts to restore profits.

1. Corporations broke their tacit social contract with union leaders and embarked on a successful multiyear campaign to roll back

wages and benefits, and in many cases to break the unions. All "sharing" of productivity gains was stopped. According to the Economic Policy Institute, between 1948 and 1979 U.S. productivity rose 108 percent and wages rose 93 percent; by contrast, between 1979 and 2013 productivity rose 65 percent, but wages rose only 8.2 percent.[35] Other studies have shown no increase in most workers' real wages since the 1970s. The unions were effectively neutralized as a limiting force on corporate power.

2. Corporations campaigned, with considerable success, to weaken or eliminate existing environmental protection laws, and to prevent adoption of any measures to reduce greenhouse gas emissions. The role played by Exxon and the petro-billionaire Koch brothers in undermining climate science has been well documented. What's also noteworthy is the absence of pushback from other giant corporations, any of which could easily have funded an effective pro-science campaign. Fossil fuels are so important to the operation of every corporation that none is willing to interfere with production, or even to offend the corporations that control the supply.

3. Corporations initiated, slowly at first and much more rapidly by the end of the 1990s, a "three-decades-long tidal wave of outsourcing of manufacturing to low-wage countries" that was "driven by capitalist firms based in the imperialist economies, impelled by their insatiable urge to cut costs by substituting relatively expensive domestic labor with cheap southern labor."[36] The largest non-financial corporations now conduct a majority of their manufacturing in foreign affiliates, and about 40 percent of world trade now results from outsourcing and subcontracting to overseas companies.[37]

Although all three measures strengthened fossil capitalism and increased greenhouse gas emissions, the third merits more extensive discussion.

Industry has moved south primarily to take advantage of much lower wages: taking into account differences in the cost of living, wages in Pakistan, Madagascar, Indonesia, and China are one-tenth of what someone doing similar work in the United States would receive. China in particular now has an economy that is "structured around the offshoring needs of multinational corporations geared to obtaining low unit labor costs by taking advantage of cheap, disciplined labor in the Global South."[38]

However, as Douglas Dowd points out, lower payroll costs are not the only reason for outsourcing production to the South:

> In the poorer parts of the world—most of Asia and Latin America, and parts of Europe—in addition to the "global labor arbitrage" . . . there is "global tax arbitrage," and what may be called "environmental arbitrage." The latter means that companies whose production destroys forests, or water supplies, or the air, whatever, may easily buy their way out of any restrictions; and on a given revenue, they may be assured of exemption from taxes—with the added attraction that the existence of such possibilities in distant places means they can often be bargained for successfully at home.[39]

Large transfers of manufacturing to low-wage countries have had the direct effect of increasing pollution in those countries, particularly greenhouse-gas emissions. Their industrial revolution, like the one in Europe and North America in the 1800s, depends on cheap power, and that has usually meant using coal, often low-quality coal, to produce electricity. As is widely known, more greenhouse gas is now produced in China than in any other country; less often noted is how much of that gas is generated to produce goods that are destined for the Global North. Rich countries have outsourced a significant part of their environmental destruction to the Global South.

And, as Andreas Malm has pointed out, available statistics do not include the emissions that result from exported production.

> The figures for exports are likely to be underestimates, since they only take into account the emissions caused directly by the production of commodities shipped overseas. They do not include the emissions from the construction of a factory, a highway to connect two industrial zones, a high-rise apartment building to accommodate workers or any other infrastructural project geared to the expansion of the export sector, nor the emissions from household consumption and other economic activities indirectly stimulated by the miracle of Chinese export. Would these be added, the figures would, of course, be dramatically higher.[40]

In *Fossil Capital*, Malm summarizes the latest major change in capitalist production:

> China had, relatively speaking, low wages and high carbon intensity, certain other countries had high wages and low carbon intensity, and capital flowed from the latter to the former.... If Manchester was the "chimney of the world" in the 1840s, the People's Republic of China assumed that position in the early twenty-first century primarily because globally mobile capital seized upon it as its workshop.[41]

Emissions in Motion

One result of the export of manufacturing to China and elsewhere in the South has been a huge increase in international shipping, and a consequent increase of emissions from ships. The bunker fuel used by large cargo ships is the cheapest and dirtiest fuel available: it's as thick as asphalt, made from the waste that's left after all other fuels have been refined from crude oil. CO_2 emissions from ships burning bunker have grown 3.7 percent a year since 1990. A large container ship burns 350 tons of fuel a day and emits more CO_2 each year than many coal-fired power stations.[42]

It's difficult to obtain figures, because international shipping, like the military, isn't included in anyone's official accounting, but it has been estimated that the world's container ships, which carry 90 percent of world trade, produce more greenhouse gas each year than 205 million cars. If cargo ships were a country, they would be the sixth-largest emitter, ranking just below Japan.[43] Shipping produces more greenhouse gases than all of Africa combined.

Bunker fuel has high sulfur content. European rules allow up to 45,000 parts per million of sulfur in marine fuel, compared to just 15 ppm in automobile fuel. High sulfur content means high levels of particulate pollution, which has a wide range of health effects, including asthma, heart attacks, and lung cancer. Scientists estimate that burning bunker causes about 87,000 premature deaths among people in coastal regions each year.[44]

Germany's Institute for Atmospheric Research predicts that under business-as-usual conditions, by 2050 carbon dioxide and sulfur dioxide emissions from ships could double, and nitrogen oxide emissions from ships will exceed all global road traffic.[45]

Oceangoing ships also carry waste from wealthy countries for disposal in the Third World, a process that only makes financial sense because the shippers and shipping companies don't pay the environmental costs. As Robert Biel writes: "Globalization as a whole is entirely premised on the fiction that the energy cost of such flows can be ignored indefinitely."[46] The Anthropocene, we might say, is Mother Earth presenting the bill.

Plastic Plague

If the Great Acceleration graphs had included one for plastics, it might have been the most dramatic of all. A report prepared in January 2016 for the billionaires who attend the annual World Economic Forum in Davos, Switzerland, provides telling data on the size and impact of this sector of the fossil economy.[47]

THE FERTILIZER FOOTPRINT

Scientists now know that the 17% increase of N_2O in the atmosphere since the preindustrial era is a direct result of chemical fertilizers, owing especially to the deployment of the so-called Green Revolution programs of the 1960s that brought chemical fertilizers into use in Asia and Latin America. They also now know that the amount of N_2O emissions resulting from the application of nitrogen fertilizers is more in the range of 3–5%, a dramatic increase from the IPCC's assumption of 1%.

Yet even this 3–5% estimate does not go far enough in assessing current and future emissions from fertilizers. First, fertilizer use is expanding fastest in the tropics, where soils generate even higher rates of N_2O emissions per kg of nitrogen applied, particularly when the soils have been deforested. Secondly, fertilizer use per hectare is growing and new studies show that the rate of N_2O emissions increases exponentially as more fertilizer is applied.

Chemical fertilizers are addictive. Because they destroy the natural nitrogen in the soils that is available to plants, farmers have to use more and more fertilizers every year to sustain yields. Over the past 40 years, the efficiency of nitrogen fertilizers has decreased by two-thirds and their consumption per hectare has increased by seven times.

From about 100,000 tonnes in 1939, plastic production soared after the war: 1.3 million metric tons in 1953; 15 million in 1964; 311 million in 2014. If current trends continue, worldwide production will double by 2035, and double again by 2050. The industry uses about six percent of global oil production, more than aviation, and that figure will increase to 20 percent by mid-century if business as usual continues.

The largest use of plastic, 26 percent of all production, is for packaging—products designed to be thrown away, made from materials that never die. Despite industry hype, only 14 percent of

The effect on organic matter, the world's most important carbon sink, is the same. Despite industry propaganda to the contrary, recent studies demonstrate that chemical fertilizers are responsible for much of the massive loss of organic matter that has occurred in the world's soils since the preindustrial era.

"In numerous publications spanning more than 100 years and a wide variety of cropping and tillage practices, we found consistent evidence of an organic carbon decline for fertilized soils throughout the world," says University of Illinois soils scientist Charlie Boast.

Soils around the world have lost, on average, at least 1–2 percentage points of organic matter in the top 30 cm since chemical fertilizers began to be used. This amounts to some 150,000–205,000 million tonnes of organic matter, which has resulted in 220,000–330,000 million tonnes of CO_2 emitted into the air or 30 per cent of the current excess CO_2 in the atmosphere!

The overall contribution of chemical fertilizers to climate change has thus been drastically underestimated and a reassessment is urgently needed. Factoring in the recent research, the growing reliance on shale gas and the impacts on soil organic matter could push estimates of the share of global GHG emissions from chemical fertilizers to as high as 10%. The world needs to move quickly to end our deadly addiction to these toxic products.

—"The Exxons of Agriculture," *GRAIN*[48]

plastic packaging is collected for recycling, and only a third of that, 5 percent of all production, is actually recycled. Fourteen percent is burned, 40 percent goes to landfills, and an appalling 32 percent enters the environment as pollution. The Davos report estimates that there are over 150 million tonnes of plastics in the oceans today, and that by 2050 plastics in the oceans will outweigh all the fish.

Further, the report says, "If the current strong growth of plastics usage continues as expected, the emission of greenhouse gases by the global plastics sector will account for 15% of the global annual carbon budget by 2050, up from 1% today."

Big Carbon Rules

Globally, there is more capital invested in oil and gas than in any other industry. Using data from the *Financial Times*, Tim Di Muzio estimates that between 2001 and 2010 the total market capitalization of the leading oil and gas producers grew an astonishing 186 percent, reaching the mind-boggling total of $3,153 trillion— remember, this was during the great recession! If state-owned oil companies are included, oil and gas capitalization totals about $6,729 trillion, more than all the world's banks combined. "The oil majors made record profits of US$655.8 billion from 2001 to 2008 while the revenue of oil exporters climbed to a total of about US$3,270 trillion from 2002 to 2008."[49]

In 2011, a United Nations report described the world's fossil fuel infrastructure:

> There are thousands of large coal mines and coal power plants, about 50,000 oilfields, a worldwide network of at least 300,000 km of oil and 500,000 km of natural gas pipelines, and 300,000 km of transmission lines. Globally, the replacement cost of the existing fossil fuel and nuclear power infrastructure is at least $15 trillion–$20 trillion. China alone added more than 300 GW [gigawatts] of coal-power capacity from 2000 to 2008, an investment of more than $300 billion, which will pay for itself only by 2030–2040 and will run maybe until 2050–2060. In fact, most energy infrastructures have recently been deployed in emerging economies and are completely new, with typical lifetimes of at least 40–60 years. Clearly, it is unlikely that the world will decide overnight to write off $15 trillion–$20 trillion in infrastructure and replace it with a renewable energy system having an even higher price tag.[50]

Despite the industry's size, it receives unparalleled financial support from governments. The International Energy Agency estimates that global subsidies to fossil fuel production and distribution totaled $548 billion in 2013; the International Monetary Fund says that if environmental damage is included in the calculation, the 2013 subsidy was $4.9 trillion.[51]

Can Capitalism De-Fossilize?

Fossil capital, says Andreas Malm, is "the energy basis of bourgeois property relations." While other materials become physically embodied in specific commodities—leather in boots, raw cotton in textiles, and so on—coal, oil, and gas are "utilized across the spectrum of commodity production as the material that sets it in physical motion." Fossil fuels are "the general lever for surplus-value production."[52] Since the early 1800s, the correlation between the growth of capitalism and the growth of greenhouse gas emissions has been so tight that Malm suggests a general law: "*Where capital goes, emissions will immediately follow. . . . The stronger global capital has become the more rampant the growth of CO_2 emissions.*"[53]

Capitalism existed before fossil fuels were introduced to the production process, and perhaps someone could write an entertaining alternative history novel about how capitalism might have developed if coal and oil and gas didn't exist, or were never discovered. But in the real world, once capitalism adopted fossil fuels there was no looking back: the two became inseparable and remain so today. Emissions continue to escalate: over 25 percent of the CO_2 added to the atmosphere since 1870 was emitted after 2000.

If it weren't for fossil capitalism, the Great Acceleration graphs would not look like hockey sticks, and Holocene conditions would not have been overwhelmed. That's obviously true for greenhouse gas emissions and surface temperature, but processes affecting the others also depend, directly or indirectly, on burning fossil fuel. Ocean acidification is caused by CO_2 emissions; artificial nitrogen cannot be made without very high temperatures and pressure; overuse of freshwater and land-system change are driven by machine-based agriculture; plastics are made from by-products of oil production, and so on. The exponential growth of fossil capitalism underlies all the changes that have driven the Earth System into the Anthropocene.

But fossil fuels also provide food, clothing, homes, heat, medicine, transportation, communications, entertainment, and much more. Is it possible to have those things without coal, oil, and gas, to decouple physical abundance from fossil fuels? Can capitalism, to invent an ugly neologism, *de-fossilize*?

If this were just a technical problem—can energy from non-fossil sources fully or even mostly replace energy from fossil fuels?—then the answer is "Probably." Credible studies from a wide range of environmental groups argue that a full transition to renewable, non-carbon fuels is physically possible. Perhaps the most comprehensive are those done by Mark Delucchi and Mark Jacobsen, who presented "A Plan to Power 100 Percent of the Planet with Renewables," in *Scientific American* in 2009, and have followed up with highly detailed technical papers in peer-reviewed journals.[54] Delucchi and Jacobson propose a 20-year project to build and install millions of wind turbines, solar installations, and other systems around the world. The estimated total cost of $100 trillion would be recovered by selling the electricity at market rates.

We do not have to endorse that particular plan, or any other proposal for that matter, to recognize that a radical reduction in fossil fuel use is *physically possible*. Nor can anyone reasonably object that the cost would be too high, in view of the much higher costs that climate change will impose on societies around the world. As proof that it could be done, advocates of such programs always cite the example of World War II, when the warring nations produced about 6,000 ships, 850,000 planes, 5 million tanks, 8 million large weapons, and the atomic bomb, in just six years.

Interviewed by the *New York Times*, Delucchi compared their plan to the Apollo moon project and the Interstate Highway System. "We really need to just decide collectively that this is the direction we want to head as a society," he said. "The biggest obstacles are social and political—what you need is the will to do it."[55] In

Scientific American, he and Jacobson cited economic and political factors as obstacles. Like many concerned scientists, they see environmental problems as fundamentally about technology: once the appropriate technical solutions have been identified, rational argument should overcome any "economic and social factors." If rational arguments don't work, it's because political will is lacking: continuing delays reflect the moral failings of politicians.

Actually, economic and political factors make all the difference. Even at the low prices that prevailed in early 2016 ($31.60 a barrel on February 1) global proven oil reserves were worth about $50 trillion. No capitalist would willingly exchange that for the chance to sell green electricity twenty years from now, let alone write off $15–$20 trillion in infrastructure, but that's what the Delucchi-Jacobson plan requires. The idea that the powerful oil industry would agree to a voluntary shutdown of its operations is ludicrous.

World War II, interstate highways, and the Apollo project show that capitalism is compatible with large-scale government-funded projects—but each of them pumped tax dollars *into* the petroleum, military, and automotive industries that the Delucchi-Jacobson plan would undermine. That plan may be good for Earth's future, but the energy status quo is essential for the profit system today, and that will always take precedence. If an environmental plan would undermine the class and power relationships that define fossil capitalism, even if it would prevent climate catastrophe, then all the rational argument in the world won't produce the political will to implement it.

Fossil fuels are not an overlay that can be peeled away from capitalism, leaving the system intact. They are embedded in every aspect of the system.

> High levels of carbon-based energy are central to virtually every productive and reproductive process within the system—not just to manufacturing industry, but to food production and distribution, the heating and functioning of office blocks, getting labor power to and from workplaces, providing

it with what it needs to replenish itself and reproduce. To break with the oil-coal economy means a massive transformation of these structures, a profound reshaping of the forces of production and the immediate relations of production that flow out of them.[56]

Carrying out that transformation is the challenge socialists face in the Anthropocene.

11

We Are Not All in This Together

We continue sailing on our *Titanic* as it tilts slowly into the darkened sea. The deckhands panic. Those with cheaper tickets have begun to be washed away. But in the banquet halls, the music plays on. The only signs of trouble are slightly slanting waiters, the kabobs and canapes sliding to one side of their silver trays, the somewhat exaggerated sloshing of the wine in the crystal wineglasses. The rich are comforted by the knowledge that the lifeboats on the deck are reserved for club-class passengers. The tragedy is that they are probably right.

—ARUNDHATI ROY[1]

It is difficult to find a liberal environmentalist who doesn't at some point insist that we are all passengers on Spaceship Earth, sharing a common fate and a common responsibility for the ship's safety. Al Gore, for example, tells us: "We all live on the same planet. We all face the same dangers and opportunities, we share the same responsibility for charting our course into the future."[2]

In reality, a handful of Spaceship Earth's passengers travel first-class, in plush air-conditioned cabins with every safety feature, including reserved seats in the very best lifeboats. The majority are on wooden benches in third-class, exposed to the elements, with no lifeboats at all.

In the twenty-first century, fossil capitalism is characterized not just by inequality—that's always been a feature of class society—but by *gross* inequality, an unparalleled accumulation of wealth in the hands of a very few, coupled with mass poverty that is enforced by all the economic, political, and military resources the ultra-rich can muster.

Many studies, articles, and reports have documented the disproportionate wealth at the top. Two statistics can stand for them all: In 2015, the richest 1 percent of the world's population owned as much as the remaining 99 percent combined, and just 62 individuals owned more than the poorest three and a half billion.[3] That obscene inequality results not just in the ultra-rich consuming a vastly excessive proportion of the world's resources, although that does happen, but in a concentration of political and economic *power* that makes a mockery of capitalism's democratic pretensions.

The combination of unequal wealth and power with the extreme weather and climate change described in chapter 6 is already having disastrous impacts on the majority of the world's people. The line is not only between rich and poor, or comfort and poverty: it is between survival and death.

The Victims of Climate Change

Climate change and extreme weather events are not devastating a random selection of human beings from all walks of life. There are no billionaires among the dead, no corporate executives living in shelters, no stockbrokers watching their children die of malnutrition. Overwhelmingly, the victims are poor and disadvantaged. Globally, 99 percent of weather disaster casualties are in developing countries, and 75 percent of them are women.[4]

The pattern repeats at every scale. Globally, the South suffers far more than the North. Within the South, the very poorest countries, mostly in Africa south of the Sahara, are hit hardest. Within each country, the poorest people—women, children, and the elderly—are most likely to lose their homes and livelihoods from climate change, and most likely to die.

The same pattern occurs in the North. Despite those countries' overall wealth, the poorest people there suffer disproportionately as climate change accelerates. Most of the people who died in the 1995 Chicago heat wave were poor and elderly, living alone in tiny apartments or rooms where air conditioning was an unaffordable luxury, in neighborhoods where opening the windows was dangerous:

> The geography of heat-wave mortality was consistent with the city's geography of segregation and inequality: eight of the ten community areas with the highest death rates were virtually all African American, with pockets of concentrated poverty and violent crime, places where old people were at risk of hunkering down at home and dying alone during the heat wave.[5]

Chronic hunger, already a severe problem in much of the world, will be made worse by climate change. As Oxfam reports: "The world's most food-insecure regions will be hit hardest of all."

> Already, food production and prices are being hit globally by extreme climate events. Other climate impacts and climate change have also been shown to be a key factor in disasters, such as the 2011 Horn of Africa drought. According to one estimate, climate change and its impacts on hunger and communicable diseases are currently responsible for 400,000 deaths a year in the world's poorest countries. The poorest people are bearing the brunt as climate change exacerbates preexisting conditions that make them more exposed to the risk of food insecurity.
>
> Today one person in eight goes to bed hungry. It has been estimated that the number of people at risk of hunger by 2050 could increase by 10–20 percent more than would be expected without climate change; and child malnutrition could increase

by 21 percent, eliminating the improvements that may otherwise have occurred.[6]

Unmitigated climate change will lock the world's poorest countries and their poorest citizens in a downward spiral, leaving
hundreds of millions facing malnutrition, water scarcity, ecological threats, and loss of livelihood. Children will be among the
primary victims, and the effects will last for lifetimes: studies in
Ethiopia, Kenya, and Niger show that being born in a drought year
increases a child's chances of being irreversibly stunted by 41 to 72
percent.[7]

Exclusion and Exterminism

Most reports and studies that document the connection between
climate change and poverty treat the issue as a tragic but inevitable case of cause and effect: poor environmental conditions cause
human suffering. That's only part of the story. Much of what seems
to be a result of climate change is actually driven by policies of
racism and exclusion that are embedded in the illogical logic of
fossil capitalism.

In the days following Hurricane Katrina, several hundred
people, including babies in strollers and adults in wheelchairs,
attempted to leave New Orleans by walking across one of the few
possible escape routes, the bridge to Gretna, a small city on the
far side of the Mississippi River. At the foot of the bridge, armed
deputies from the Gretna police force blocked their way and fired
shotguns in the air, forcing them back. Two Emergency Medical
Service workers from San Francisco, who had been trapped in the
city after a paramedics convention, described the scene:

> The sheriffs informed us that there were no buses waiting. We
> questioned why we couldn't cross the bridge anyway, espe-
> cially as there was little traffic on the six-lane highway. They
> responded that the west bank was not going to become New

Orleans, and there would be no Superdomes in their city. These were code words for: if you are poor and Black, you are not crossing the Mississippi River, and you are not getting out of New Orleans.[8]

That episode was just a small part of the "savage sorting" (Sakia Sassen's phrase) that took place in New Orleans in 2005. Chester Hartman and Gregory Squires, who titled their account *There Is No Such Thing as a Natural Disaster*, write:

> Those with means left when they knew the storm was coming: They had access to personal transportation or plane and train fare, money for temporary housing, in some cases second homes. Guests trapped in one luxury New Orleans hotel were saved when that chain hired a fleet of buses to get them out. Patients in one hospital were saved when a doctor who knew Al Gore contacted the former Vice President, who was able to cut through government red tape and charter two planes that took them to safety. This is what is meant by the catchphrase "social capital"—a resource most unevenly distributed by class and race. . . .
>
> It should have been no surprise when Katrina hit New Orleans that the areas damaged were 45.8% black, compared to 26.4% in undamaged areas, and that 20.9% of the house-holds in damaged areas were poor, compared to 15.3% in undamaged areas.[9]

Every account of Katrina confirms that picture, and a similar savage sorting took place when Hurricane Sandy hit New York in 2012. If such extreme divisions occur in the world's richest country, we know that it is far, far worse in places where absolute poverty is the norm.

In 1980, the English historian and antiwar activist E. P. Thompson proposed the word *exterminism* for "those charac-teristics of a society—expressed, in differing degrees, within its economy, its polity and its ideology—which thrust it in a direction whose outcome must be the extermination of multitudes."[10] Thompson's focus was on the potential results of the nuclear arms race, but others have extended the concept to address the impact of systemic ecological destruction on human beings and society. The best definition was given by Stan Goff in his summing-up of the lessons of Katrina:

> [Exterminism is] the tacit or open acceptance of the necessity for mass exterminations or die-offs (*often beginning with mass displacements*) as the price for continued accumulation and the political dominance of a ruling class. . . .
>
> Exterminism is not totally, or even most often, characterized by offensive action against whole populations, but frequently accomplished by calculated neglect—the instruments of which are poverty, disease, malnutrition, and "natural" disasters . . . and frequently facilitated by economic isolation and the mass displacement of populations.[11]

We could see exterminism in action in 2014 and 2015, when thousands of people from the Middle East and Africa drowned while attempting to reach Europe. They were part of a mass exodus triggered by fossil capitalism: by the worst droughts and highest temperatures ever recorded, and by the brutal wars rooted in the efforts of the United States, Canada, and Western Europe to protect access to oil. Explaining why her government refused to support a sea rescue program that could save refugees from death at sea, UK Foreign Office minister, Lady Anelay, told the House of Lords that such a program would mean "encouraging more migrants to attempt the dangerous sea crossing and thereby leading to more tragic and unnecessary deaths":

> We understand that by withdrawing this rescue cover we will be leaving innocent children, women and men to drown who we would otherwise have saved. But eventually word will get

around the war-torn communities of Syria and Libya and the other unstable nations of the region that we are indeed leaving innocent children, women and men to drown. And when it does, they will think twice about making the journey. And so eventually, over time, more lives will be saved.[12]

That might be called "calculated neglect," but it would not be unreasonable to call it murder. It's certainly clear that capitalist politicians, including the many who ostentatiously flaunt their supposed Christianity, have no intention of emulating Christ's Good Samaritan.

Environmental Militarism

Governments that follow such policies say that they want to help people adapt so they can stay in their home countries, but their actions belie their words. A case in point is the Green Climate Fund, set up in 2010 to provide $100 billion a year to assist Third World nations in adapting to climate change. Five years later, concrete pledges to the fund totaled just $10.2 billion, and less than one billion had actually been contributed. As India's representative on the GCF Board said, "At this pace we will not be able to do anything much."[13]

That's not to say the rich countries aren't spending money to deal with climate change in the Third World—they're just spending it in other ways. As Christian Parenti says, "The anticipation of increased conflict in a world remade by climate change has led the militaries of the Global North toward an embrace of militarized adaptation."[14]

What we might call "environmental militarism" emerged in the years after the fall of the Soviet Union, when the U.S. military was looking for reasons to keep its budget as large as possible. One such reason came in an influential book by Robert Kaplan, *The Coming Anarchy*. In a chapter titled "The Environment as a Hostile Power," Kaplan identified a new enemy.

It is time to understand "the environment" for what it is: *the* national-security issue of the early twenty-first century. The political and strategic impact of surging populations, spreading disease, deforestation and soil erosion, water depletion, air pollution, and, possibly, rising sea levels in critical, overcrowded regions like the Nile Delta and Bangladesh— developments that will prompt mass migrations and, in turn, incite group conflicts—will be the core foreign-policy challenge from which most others will ultimately emanate, arousing the public and uniting assorted interests left over from the Cold War.

The environment, Kaplan said, "will define a new threat to our security . . . allowing a post–Cold War foreign policy to emerge inexorably by need rather than by design."[15]

Bill Clinton's National Security Strategy took the same view, listing "environmental degradation, natural resource depletion, rapid population growth and refugee flows" as security risks, in the same sentence as terrorism and drug trafficking. "An emerging class of transnational environmental issues are increasingly affecting international stability and consequently will present new challenges to U.S. strategy."[16]

Taking that policy further in the Bush Jr. era, in 2003 the Pentagon commissioned a study titled *An Abrupt Climate Change Scenario and Its Implications for United States National Security.* The authors, consultants with the high-powered Global Business Network, argued that rapid climate change "could potentially destabilize the geo-political environment, leading to skirmishes, battles, and even war":

> Nations with the resources to do so may build virtual fortresses around their countries, preserving resources for themselves. Less fortunate nations, especially those with ancient enmities with their neighbors, may initiate struggles for access to food, clean water, or energy.

They left no doubt about who would be who in that scenario:

The United States and Australia are likely to build defensive fortresses around their countries because they have the resources and reserves to achieve self-sufficiency. With diverse growing climates, wealth, technology, and abundant resources, the United States could likely survive shortened growing cycles and harsh weather conditions without catastrophic losses. Borders will be strengthened around the country to hold back unwanted starving immigrants from the Caribbean islands (an especially severe problem), Mexico, and South America. Energy supply will be shored up through expensive (economically, politically, and morally) alternatives such as nuclear, renewables, hydrogen, and Middle Eastern contracts.[17]

Let's be crystal clear: this was a call for the use of armed force against starving people. This is precisely the policy advocated decades ago by the right-wing eugenicist and overpopulation ideologue Garrett Hardin, in his notorious 1974 article, "Lifeboat Ethics: The Case Against Helping the Poor." He wrote: "In a less than perfect world, the allocation of rights based on territory must be defended. . . . It is unlikely that civilization and dignity can survive everywhere; but better in a few places than in none."[18]

Environmentalist Barry Commoner replied that such policies have *nothing* in common with civilization and dignity.

Here, only faintly masked, is barbarism. It denies the equal right of all the human inhabitants of the earth to a humane life. It would condemn most of the people of the world to the material level of the barbarian, and the rest, the "fortunate minorities," to the moral level of the barbarian. Neither within Hardin's tiny enclaves of "civilization," nor in the larger world around them, would anything that we seek to preserve—the dignity and the humaneness of man, the grace of civilization—survive.[19]

The 2003 report caused an uproar when it was leaked, leading Pentagon officials to insist that the scenario was speculative, but they didn't abjure the "virtual fortress" response to climate crisis. Many examples, including the walls and armed patrols on the U.S.-Mexico border, Australia's internment of refugees in brutal island prison camps, Britain's exclusion of refugees camped in Calais, and Hungary's wall against Syrian refugees, show that the report's principal error was its assumption that refugee exclusion policies would only happen in case of a sudden global climate shift. What Parenti calls the *politics of the armed lifeboat*—"responding to climate change by arming, excluding, forgetting, repressing, policing and killing"[20]—is a major element of the climate-change policies of wealthy countries today. It is certainly the best-financed part.

U.S. officials routinely describe climate change not as a matter of justice but as a "threat multiplier" that must be met with force. Political scientist Robyn Eckersley writes that while U.S. climate negotiators refuse to commit to any practical action that might reduce greenhouse gas emissions, its military officials actively treat climate change as a security problem caused by others: "Environmental threats are something that foreigners do to Americans or to American territory."[21]

Environmental Apartheid

In 1844, Frederick Engels described how the streets of Manchester were carefully laid out so that the "money aristocracy can take the shortest road through the middle of all the labouring districts to their places of business, without ever seeing that they are in the midst of the grimy misery that lurks to the right and the left."[22] Today, that physical separation is global. What Archbishop Tutu calls "adaptation apartheid" is business as usual in the Anthropocene. [23]

While the military targets climate-change victims as enemies of the capitalist way of life, global elites are preparing for dark times by creating protected spaces for themselves, their families, and their servants in the hope of ensuring that they continue to get more than their share of the world's wealth, no matter what happens to anyone else.

Mike Davis, noting that climate change will have "dramatically unequal impacts across regions and social classes, inflicting the greatest damage upon poor countries with the fewest resources for meaningful adaptation," warns that anyone who expects solutions in the present social order must depend on "the transmutation of the self-interest of rich countries and classes into an enlightened 'solidarity' with little precedent in history":

> Instead of galvanizing heroic innovation and international cooperation, growing environmental and socio-economic turbulence may simply drive elite publics into more frenzied attempts to wall themselves off from the rest of humanity. Global mitigation, in this unexplored but not improbable scenario, would be tacitly abandoned—as, to some extent, it already has been—in favor of accelerated investment in selective adaptation for Earth's first-class passengers. The goal would be the creation of green and gated oases of permanent affluence on an otherwise stricken planet.[24]

In *Evil Paradises: Dreamworlds of Imperialism*, Davis and co-author Daniel Bertrand Monk document the "unprecedented spatial and moral secession of the wealthy from the rest of humanity," in custom-built communities of unspeakable luxury:

> Modern wealth and luxury consumption are more enwalled and socially enclaved than at any time since the 1890s. . . .
> On a planet where more than 2 billion people subsist on two dollars or less a day, these dreamworlds enflame desires—for infinite consumption, total social exclusion and physical security, and architectural monumentality—that are clearly incompatible with the ecological and moral survival of humanity. . . .

They expand our understanding of what Luxemburg and Trotsky had in mind when they warned of "Socialism or Barbarism.". . . They are willful, narcissistic withdrawals from the tragedies overtaking the planet. The rich will simply hide out in their castles and television sets, desperately trying to consume all the good things of the earth in their lifetimes.[25]

Private jets and super-yachts enable the ultra-rich to flee to the remote dreamworlds that Davis and Monk describe, but even the not-quite-that-rich can shield themselves from Earth's attacks, while remaining in the crumbling cities of the North. Naomi Klein writes:

Sparked by events like Superstorm Sandy, new luxury real estate developments are marketing their gold-plated private disaster infrastructure to would-be residents—everything from emergency lighting to natural gas–powered pumps and generators to thirteen-foot floodgates and watertight rooms sealed "submarine-style," in the case of a new Manhattan condominium. As Stephen G. Kliegerman, the executive director of development marketing for Halstead Property, told the *New York Times*: "I think buyers would happily pay to be relatively reassured they wouldn't be terribly inconvenienced in case of a natural disaster."[26]

Meanwhile, the storm left thousands of New York's poor trapped in their public housing apartments for as long as three weeks, without electricity, heat, or water. "Terribly inconvenienced" doesn't begin to describe it.

"Capitalist accumulation," Marx wrote, "constantly produces . . . a population which is superfluous to capital's average requirements for its own valorization, and is therefore a surplus population." As capital expands, "the entanglement of all peoples in the net of the world market" creates an ever-growing global divide between rich and poor. "Along with the constant decrease in the number of capitalist magnates, who usurp and monopolize all the advantages of this process of transformation, the mass of misery, oppression, slavery, degradation and exploitation grows."[27]

What we see today goes beyond the horrors Marx described. As capitalism has plundered the world, it has made an increasingly large proportion of the population not just "relatively redundant" but *absolutely surplus* to capital's profit-making requirements. They aren't needed as producers or consumers, and few of them ever will be. So, as David Harvey writes, they can be abandoned:

> Deaths from starvation of exposed and vulnerable populations and massive habitat destruction will not necessarily trouble capital (unless it provokes rebellion and revolution) precisely because much of the world's population has become redundant and disposable anyway. And capital has never shrunk from destroying people in pursuit of profit.[28]

Hundreds of millions have already been pushed to the outer edges of the global economy and beyond, denied access to the minimum requirements of life, and left to survive the deteriorating global environment on their own. Excluded from the fossil economy, they have become its primary victims. The insane logic of exterminism and apartheid rules.

If this continues, if fossil capitalism remains dominant, the Anthropocene will be a new dark age of barbarous rule by a few and barbaric suffering for most. That's why the masthead of the web journal *Climate & Capitalism* carries a slogan adapted from Rosa Luxemburg's famous call for resistance to impending disaster in the First World War: "Ecosocialism or barbarism: There is no third way."

Indeed, as István Mészáros argues, the choice may be even more stark:

If I had to modify Rosa Luxemburg's dramatic words, in relation to the dangers we now face, I would add to "socialism or barbarism" this qualification: *"barbarism if we are lucky."* For the extermination of humanity is the ultimate concomitant of capital's destructive course of development.[29]

PART THREE

THE ALTERNATIVE

When Gus Speth included the Great Acceleration graphs in his book *The Bridge at the Edge of the World*, he gave a different name to the process they display: "If we could speed up time, it would seem as if the global economy is crashing against the earth—the Great Collision."[1]

To fully appreciate that statement, it's important to know that Speth has spent most of his adult life trying to save the environment by working inside the system. He was a senior environmental advisor to President Jimmy Carter and later to Bill Clinton. In the 1990s he was administrator of the United Nations Development Program and chair of the United Nations Development Group. *Time* magazine called him "the ultimate insider."[2] So it's significant that after forty years he has concluded that working inside the system failed because the capitalist system itself is the cause of environmental destruction:

> These features of capitalism, as it is constituted today, work together to produce an economic and political reality that is highly destructive of the environment. An unquestioning society-wide commitment to economic growth at almost any cost; enormous investment in technologies designed

with little regard for the environment; powerful corporate interests whose overriding objective is to grow by generating profit, including profit from avoiding the environmental costs they create; markets that systematically fail to recognize environmental costs unless corrected by government; government that is subservient to corporate interests and the growth imperative; rampant consumerism spurred by a worshipping of novelty and by sophisticated advertising; economic activity so large in scale that its impacts alter the fundamental biophysical operations of the planet; all combine to deliver an ever-growing world economy that is undermining the planet's ability to sustain life.[3]

The fact that this critique comes from someone who has worked for so long inside the system makes it credible and powerful. Those hockey-stick Great Acceleration graphs, now heading nearly straight up, display critical impacts of aggressive fossil capitalism spreading across the world to satisfy its appetite for capital accumulation and burning ever greater quantities of fossil fuels to achieve that end. Every day, capitalism's relentless drives accelerate, the trend lines continue to rise, and the crisis becomes more severe.

This is, by any reasonable measure, a planetary emergency. As leading climate scientist James Hansen says, "Planet Earth, creation, the world in which civilization developed, the world with climate patterns that we know and stable shorelines, is in imminent peril."[4] By the end of this century, the United Nations Development Program warns, "the spectre of catastrophic ecological impacts could have moved from the bounds of the possible to the probable."[5]

The Anthropocene, write three leading observers, "heralds a new geological regime of existence for the Earth and a new human condition":

Living in the Anthropocene means living in an atmosphere altered by the 575 billion tonnes of carbon emitted as carbon dioxide by human activities since 1870. It means inhabiting an impoverished and artificialized biosphere in a hotter world increasingly characterized by catastrophic events and new risks, including the possibility of an ice-free planet. It means rising and more acidic seas, an unruly climate and its cortege of new and unequal sufferings. It's a world where the geographical distribution of population on the planet would come under great stress.[6]

Capitalism has driven the Earth System to a crisis point in the relationship between humanity and the rest of nature. If business as usual continues, the first full century of the Anthropocene will be marked by rapid deterioration of our physical, social, and economic environment. The decay of the biosphere will be most noticed as global warming and extreme weather, but we can also expect rising ocean levels leading to widespread flooding, the collapse of major fisheries, poisoned rivers, and more. Every planetary boundary is threatened, and a catastrophic convergence of multiple Earth System failures is possible.

If that happens, the Anthropocene may be the shortest of all epochs, a transition period from the Holocene to something far worse.

The only way to avoid that is with methods that are anathema to capitalism. Profit must be removed from consideration; all changes must be made as part of a democratically created and legally binding global plan that governs both the conversion to renewables and the rapid elimination of industries and activities, such as arms production, advertising, and factory farming, that only produce what John Ruskin called "illth," the opposite of wealth.

The physical changes are serious, but they will not, by themselves, determine what life will be like in the Anthropocene—only human action (or inaction) can do that. The world our children and grandchildren will inherit will be defined by the way our generation responds to the planetary emergency.

12

Ecosocialism and Human Solidarity

Even an entire society, a nation, or all simultaneously existing societies taken together are not owners of the earth, they are simply its possessors, its beneficiaries, and have to bequeath it in an improved state to succeeding generations, as good heads of households.

—KARL MARX[1]

If any attempt to change society, and not just mend it, is branded angrily and contemptuously as utopian, then, turning the insult into a badge of honor, we must proudly proclaim that we are all utopians.

—DANIEL SINGER[2]

In New York in 2012, amid the devastation and suffering caused by Hurricane Sandy, something remarkable happened. While federal, state, and city officials dithered, thousands of people joined Occupy Sandy, a self-organized volunteer relief campaign that provided food, clothing, and essential support in the poorest, most damaged parts of the city. At peak, some 60,000 volunteers were working out of ten hubs, in a project that emphasized mutual aid rather than charity—a project that actually met people's needs.

For those who see humans as motivated by greed and self-interest, such an outpouring of human solidarity is incomprehensible, but as Rebecca Solnit showed in *A Paradise Built In Hell*, what happened after Sandy has happened many times before. In New Orleans in 2005, for example, "thousands of people survived Hurricane Katrina because grandsons or aunts or neighbors or complete strangers reached out to those in need all through the Gulf Coast and because an armada of boat owners from the surrounding communities and as far away as Texas went into New Orleans to pull stranded people to safety."[3]

Unlike "the minority in power, who often act savagely in a disaster," Solnit shows,

> in the wake of an earthquake, a bombing, or a major storm, most people are altruistic, urgently engaged in caring for themselves and those around them, strangers and neighbors as well as friends and loved ones. The image of the selfish, panicky, or regressively savage human being in times of disaster has little truth to it. Decades of meticulous sociological research on behavior in disasters, from the bombings of World War II to floods, tornadoes, earthquakes, and storms across the continent and around the world, have demonstrated this. . . .
>
> The positive emotions that arise in those unpromising circumstances demonstrate that social ties and meaningful work are deeply desired, readily improvised, and intensely rewarding.

In normal times, "the very structure of our economy and society prevents these goals from being achieved." But these "ubiquitous fleeting utopias that are neither coerced nor countercultural but universal" show that another world is possible.[4] They prefigure the "solidarian society" that Michael Lebowitz describes in *The Socialist Alternative*.

> Building a solidarian society means going beyond our own particular interests—*or, more accurately, understanding that our particular interest is that we live in a society in which everyone has the right to full human development*. It means that our premise is the concept of a human community.[5]

The possibility of building such a society offers a reason for hope in the Anthropocene.

The Problem of Time

How long do we have? How soon must emissions be drastically reduced, to avoid dangerous climate change?

The answer, in one sense, is that it is already too late. Dangerous climate change is already upon us. Even if we stop all emissions today, things are going to get worse, because warming depends on the total amount of greenhouse gas in the atmosphere, and it takes years for today's emissions to have their full effect. What's more, the natural processes that remove excess CO_2 from the atmosphere take centuries, even millennia, to do their work. More melted glaciers and ice sheets, more rising oceans, and more extreme weather, are inevitable.

A better question is, as Kevin Anderson of Britain's Tyndall Centre puts it, when will we cross the line between *dangerous* climate change and *extremely dangerous* climate change? Of course, that depends on what we mean by "extremely," but if we agree that the global temperature increase should be kept below 1.5° or 2°C, then the answer is: *not long*. To prevent a 2°C (3.6°F) increase in this century, most models require drastic emission reductions no later than 2020, and even then, after 2050 most require "negative emissions"—removal of CO_2 from the atmosphere using some unknown technology. (In fact, as Anderson has pointed out, many of the IPCC's models require CO_2 concentrations to start declining in 2010: unless someone invents time travel, those models have already failed.)

So we don't have much time. And given the refusal of our rulers to act—witness the failure of every UN climate meeting for two decades to adopt any concrete measures against fossil fuels—it is now unlikely that the necessary changes will be made in time to stop a 2-degree increase.

If such a boundary passes, environmentalists and socialists must of course continue to fight, to provide aid and refuge to the victims of climate change, and to prevent the destruction from going further. But as the destruction increases, the barriers that prevent sustainable human development get higher and higher. As the Brazilian ecosocialist and atmospheric scientist Alexandre Costa writes, "The fight to avoid a catastrophic outcome to this crisis engendered by capitalism is the fight to safeguard the material conditions for survival with dignity of humankind. . . . Socialism is not possible on a scorched Earth."[6]

We do not know how long we have, but we do know that the fight simply cannot wait. And we know that just fighting isn't enough: to succeed, we must simultaneously work for immediate changes *and* advance a vision of the world we want to build. If we are serious about changing the world, we must recognize and resolve what the late Daniel Singer described as "the dilemma that has faced all socialist parties and, really, all movements not resigned to run the world as it is":

> The problem is that they must struggle within the rugged reality of existing society and provide solutions that, sooner or later, lead beyond the confines of that society. If they limit themselves to the questions connected with the future . . . they will find themselves miles ahead of the movement in splendid sectarian isolation. Yet, if they become bogged down in daily battles and ignore the future, they will forget that their original purpose was to reshape society in order to alter the fate of working people. . . . The real question . . . is how to reconcile the two, how to defend the interests of working people within the existing society and build up this struggle into a general offensive challenging the very foundations of the system. [7]

István Mészáros has made the same point, more firmly and concisely: "Without identifying the *overall destination* of the journey, together with the *strategic direction* and the necessary *compass* adopted for reaching it, there can be no hope of success."[8]

Destination: Ecological Civilization

Our goal is to build an *ecological civilization*, a society that, as Fred Magdoff writes, "will need to be the opposite of capitalism in essentially all aspects":

> Capitalism is incompatible with a truly ecological civilization because it is a system that must continually expand, promoting consumption beyond human needs, while ignoring the limits of nonrenewable resources (the tap) and the earth's waste assimilation capacity (the sink). As a system of possessive individualism it necessarily promotes greed, individualism, competitiveness, selfishness, and an *après moi le déluge* philosophy. Engels suggested that "real human freedom" can be achieved only in a society that exists "in harmony with the laws of nature."
>
> Although it is impossible to know what future civilizations will be like, we can at least outline characteristics of a just and ecologically based society. As a system changes, it is the history of the country and the process of the struggle that bring about a new reality. However, in order to be ecologically sound, a civilization must develop a new culture and ideology based on fundamental principles such as substantive equality. It must (1) provide a decent human existence for everyone: food, clean water, sanitation, health care, housing, clothing, education, and cultural and recreational possibilities; (2) eliminate the domination or control of humans by others; (3) develop worker and community control of factories, farms, and other workplaces; (4) promote easy recall of elected personnel; and (5) re-create the unity between humans and natural systems in

all aspects of life, including agriculture, industry, transportation, and living conditions....

It would: (1) stop growing when basic human needs are satisfied; (2) not entice people to consume more and more; (3) protect natural life-support systems and respect the fact that we do not have unlimited natural resources, taking into account needs of future generations; (4) make decisions based on long-term societal/ecological needs, while not neglecting short-term needs of people; (5) run as much as possible on current (including recent past) energy instead of fossil fuels; (6) foster the human characteristics and a culture of cooperation, sharing, reciprocity, and responsibility to neighbors and community; (7) make possible the full development of human potential; and (8) promote truly democratic political and economic decision making for local, regional, and multiregional needs.[9]

Our generation may not see that vision fully accomplished, but we can get to the starting point described by one of the pioneers of revolutionary socialism and environmentalism, the British poet and artist William Morris:

The first real victory of the Social Revolution will be the establishment not indeed of a complete system of communism in a day, which is absurd, but of a revolutionary administration whose definite and conscious aim will be to prepare and further, in all available ways, human life for such a system.[10]

In *Too Many People?*, Simon Butler and I expressed that idea this way: "In every country, we need governments that break with the existing order, that are answerable only to working people, farmers, the poor, indigenous communities, and immigrants—in a word, to the *victims* of ecocidal capitalism, not its beneficiaries

---

MARXISM vs. PRODUCTIVISM

We socialists and Marxists do not share the irresponsible "productivist" credo of the 1950s and 1960s. Many social criticisms of that credo are amply justified.

One has not necessarily to accept the predictions of unavoidable absolute scarcity of energy and raw materials of the Club of Rome type in order to understand that there is a collective responsibility for the present generation of humanity to transmit to future generations an environment and a stock of natural wealth that constitute the necessary precondition for the survival and flowering of human civilization.

Neither has one to accept the impoverishing implications of permanent asceticism and austerity, so alien to the basic spirit of Marxism, which is one of enjoyment of life and infinite enrichment of human potentialities, in order to understand that the endlessly growing output of an endless variety of more and more useless commodities (increasingly, outright harmful commodities, harmful both to the environment and to the healthy development of the individual) does not correspond to a socialist ideal. Such an output simply expresses the needs and greeds of capital to realize bigger and bigger amounts of surplus value, embodied in an endlessly growing mountain of commodities.

But the rejection of the capitalist consumption pattern, combined with a no less resolute rejection of capitalist technology, should base

---

and representatives." We suggested some of the first measures such governments might take:

- Rapidly phasing out fossil fuels and biofuels, replacing them with clean energy sources such as wind, geothermal, wave, and above all, solar power;
- Actively supporting farmers to convert to ecological agriculture; defending local food production and distribution;

itself from a socialist point of view on a vigorous struggle for alternative technologies that will extend, not restrict, the emancipatory potential of machinery (i.e., the possibility of freeing all human beings from the burden of mechanical, mutilating, non-creative labor, of facilitating rich development of the human personality for all individuals on the basis of satisfaction of all their basic material needs).

We are convinced that once that satisfaction is assured in a society where the incentives for personal enrichment, greed, and competitive behavior are withering away, further "growth" will be centered around needs of "nonmaterial" production (i.e., the development of richer social relations). Moral and psychological needs will supersede the tendency to acquire and accumulate more material goods.

However "impopular" these beliefs may appear in the light of present-day fashions, we believe in the growing capacities of human intelligence, human science, human progress, human self-realization (including self-control), and human freedom, without in any way subordinating the defense of such freedoms (in the first place, freedom from want, but also freedom of thought, of creation, of political and social action) to any paternalistic instance supposedly capable of securing them for mankind.

—ERNEST MANDEL[11]

working actively to restore soil fertility while eliminating factory farms and polluting agribusinesses;

- Introducing free and efficient public transport networks, and implementing urban planning policies that radically reduce the need for private trucks and cars;
- Restructuring existing extraction, production, and distribution systems to eliminate waste, planned obsolescence, pollution, and manipulative advertising, placing industries under public

control when necessary, and providing full retraining to all affected workers and communities;

- Retrofitting existing homes and buildings for energy efficiency, and establishing strict guidelines for green architecture in all new structures;
- Ceasing all military operations at home and elsewhere; transforming the armed forces into voluntary teams charged with restoring ecosystems and assisting the victims of floods, rising oceans and other environmental disasters;
- Ensuring universal availability of high-quality health services, including birth control and abortion;
- Launching extensive reforestation, carbon farming, and biodiversity programs.[12]

Many other measures could be proposed. The exact steps such governments will take will depend on the circumstances in which they come to power. That will include, of course, matters specific to the countries they operate in, such as the economic situation, the strength of reactionary opposition forces, and so on. Their ability to act, and the ecological measures they give priority to, will also depend on the amount and extent of the environmental damage that has already occurred locally and globally. The longer we take to get rid of this destructive system, the longer it will take for humanity to deal with its consequences.

The transformation will require new knowledge and new science. New projects of the scope and scale of the International Geosphere-Biosphere Program will be needed to provide a solid scientific basis for making decisions, above all to ensure that efforts to restore the Earth System to health do not inadvertently cause more damage.

Destination: Global Human Solidarity

As a guide for deciding what actions should or should not be taken, such a revolutionary administration could refer to the partial

Charter for Human Development proposed by Michael Lebowitz:

1. Everyone has the right to share in the social heritage of human beings—an equal right to the use and benefits of the products of the social brain and the social hand—in order to be able to develop his or her full potential.
2. Everyone has the right to be able to develop their full potential and capacities through democracy, participation, and protagonism in the workplace and society—a process in which these subjects of activity have the precondition of the health and education that permit them to make full use of this opportunity.
3. Everyone has the right to live in a society in which human beings and nature can be nurtured—a society in which we can develop our full potential in communities based upon cooperation and solidarity.[13]

In the North in particular, this Charter should be viewed not only as a guide for domestic policy but as an obligatory framework for relations with the countries and peoples of the South. While victories by green-left forces in the South are important, and indeed are more likely in the near term than in the North, their ability to slow global environmental destruction will be limited. Stopping capitalist ecocide on a global scale will require governments in the Global North with the resources and the will to work toward global ecological restoration. Such governments will be able, and should be expected, to accept global responsibilities, and to devote a large proportion of their countries' resources to worldwide ecological recovery. Cuba, which provides more medical personnel and support to other countries than all the G8 countries combined, has set an example of global human solidarity that richer countries must emulate on a far larger scale.

In particular, the wealthy nations must concretely support, in Kolya Abramsky's words, "cheap (or free) and reliable sources of efficient, safe, and clean energy as a fundamental human right, not a privilege or a service."[14] Only universal access to energy based on renewable sources can *begin* to overcome the gross injustices and inequality that fossil capitalism bequeaths us. Until and unless the people of the North aid in that global transition, we will have no justification, no right at all, to object if people in the South choose to improve their standard of living using whatever fuels and technologies are available to them.

We should not delude ourselves that global environmental recovery will happen easily or quickly. To cite just one example, the United Nations has estimated that it will take thirty years to clean up the devastation caused by Shell Oil in the Ogoni people's homeland in the Niger Delta, an area of just 386 square miles. The Niger Delta is a particularly horrible example of capitalism's ecocidal role, but there are many more examples around the world, enough to dash any hope for an easy turnaround.

Destination: Ecosocialism

As Fred Magdoff says, capitalism is incompatible with a truly ecological civilization. Such a civilization can only be a socialist society, in which the economy is organized to meet social needs, not to create private profit, and in which real power rests with the great majority, not with a handful of ultra-rich individuals and giant corporations.

The word *ecosocialism*, which entered English from German in about 1980, is used today by activists who agree that there can be no true ecological revolution that is not socialist and no true socialist revolution that is not ecological. The ecosocialist movement is far from monolithic, but most of its partisans would agree that an ecosocialist society would be based on two fundamental and indivisible characteristics:

- It will be socialist, committed to democracy, to radical egalitarianism, and to social justice. It will be based on collective ownership of the means of production, and it will work actively to eliminate exploitation, profit, and accumulation as the driving forces of our economy.

- It will be based on the best ecological principles, giving top priority to stopping anti-environmental practices, to restoring damaged ecosystems, and to reestablishing agriculture and industry on ecologically sound principles.

To provide some insight into what that might mean, Joel Kovel, Michael Löwy, Danielle Follett, and I wrote the Belém Ecosocialist Declaration in 2008. It was endorsed by ecosocialists from about forty countries, so it is the closest thing there is to a global consensus statement of ecosocialist views. The following is its concluding section.[15]

·ψ·

The Ecosocialist Alternative

The ecosocialist movement aims to stop and to reverse the disastrous process of global warming in particular and of capitalist ecocide in general, and to construct a radical and practical alternative to the capitalist system. Ecosocialism is grounded in a transformed economy founded on the non-monetary values of social justice and ecological balance. It criticizes both capitalist "market ecology" and productivist socialism, which ignored the earth's equilibrium and limits. It redefines the path and goal of socialism within an ecological and democratic framework.

Ecosocialism involves a revolutionary social transformation, which will imply the limitation of growth and the transformation of needs by a profound shift away from quantitative and toward qualitative economic criteria, an emphasis on use-value instead of exchange-value.

These aims require both democratic decisionmaking in the economic sphere, enabling society to collectively define its goals of investment and production, and the collectivization of the means

SYSTEM CHANGE NOT CLIMATE CHANGE

A plethora of blueprints for an ecologically sustainable world fail, not because their proposals for a rapid conversion to renewable energy and the rational reorganization of production and consumption are far-fetched, but because they do not accept that capitalism is incapable of bringing them into being.

A socialist society run by and for the associated producers, as Marx described working people, would allow the controlling levers of the treadmill to be seized, bringing it to a halt so we can all get off and begin to rationally plan the best way forward.

Global direct military spending is running at more than US$1 trillion a year, of which the United States accounts for almost 50 percent. When related spending is factored in, U.S. military spending is above $900 billion. Just a fraction of that could eliminate starvation and malnutrition globally, provide education for every child, access to water and sanitation, and reverse the spread of AIDS and malaria. It would also enable the massive transfer of new and clean technologies to the Third World to allow poor countries to skip the dirty industrial stage of development.

The end of capitalist domination would also end the plunder of the Third World, and genuine development could ensue. With the cancellation of Third World debt, the poor countries would retain vast sums to kick-start clean development.

The wealth of the former capitalist class would also provide immense resources. According to research by Oxfam, the richest 1 percent of the world's population own more of the world's wealth than everyone else put together: a society whose goal is substantive equality will use those fortunes to build a better world for all.

of production. Only collective decisionmaking and ownership of production can offer the longer-term perspective that is necessary for the balance and sustainability of our social and natural systems.

The rejection of productivism and the shift away from quantitative and toward qualitative economic criteria involve rethinking the nature and goals of production and economic activity in general.

Genuinely democratic socialist planning could collectively redirect society's wealth into research and development of existing and new technologies to meet society's needs, while operating well within the environment's capacity to absorb the waste. We could rapidly bring forward the expansion of renewable energy, and speedily phase out coal and nuclear power stations.

With a huge boost to socially directed investment in research and development, reliable solar energy and wind power could soon become much cheaper than traditional sources. We could begin to harness the Sun's energy, which every day delivers to the Earth 17,000 times as much energy as the entire population uses.

Capitalism's dependence on the private car and truck would begin to be reversed with the rapid proliferation of mass, free public transport systems. In time, cities will no longer be designed around the private car, but around residential, community, and work hubs linked by efficient public transport.

In a society that is organized to work together to produce enough to comfortably ensure people's physical and mental well-being and social security, and in which technological advances benefit everybody without costing the environment, a new social definition of wealth will evolve.

In the words of Marx and Engels, it will be defined by the degree to which it provides the means for "all members of society to develop, maintain and exert their capacities in all possible directions," so that "the old bourgeois society, with its classes and class antagonisms," is replaced "by an association in which the free development of each is a condition of the free development of all."

—TERRY TOWNSEND[16]

Essential creative, non-productive, and reproductive human activities, such as householding, child-rearing and care, child and adult education, and the arts, will be key values in an ecosocialist economy.

Clean air and water and fertile soil, as well as universal access to chemical-free food and renewable, non-polluting energy sources, are basic human and natural rights defended by ecosocialism. Far from being "despotic," collective policy-making on the local, regional, national, and international levels amounts to society's exercise of communal freedom and responsibility. This freedom of decision constitutes a liberation from the alienating economic "laws" of the growth-oriented capitalist system.

To avoid global warming and other dangers threatening human and ecological survival, entire sectors of industry and agriculture must be suppressed, reduced, or restructured, and others must be developed, while providing full employment for all. Such a radical transformation is impossible without collective control of the means of production and democratic planning of production and exchange. Democratic decisions on investment and technological development must replace control by capitalist enterprises, investors, and banks in order to serve the long-term horizon of society's and nature's common good.

The most oppressed elements of human society, the poor and indigenous peoples, must take full part in the ecosocialist revolution in order to revitalize ecologically sustainable traditions and give voice to those whom the capitalist system cannot hear. Because the peoples of the Global South and the poor in general are the first victims of capitalist destruction, their struggles and demands will help define the contours of the ecologically and economically sustainable society in creation. Similarly, gender equality is integral to ecosocialism, and women's movements have been among the most active and vocal opponents of capitalist oppression. Other potential agents of ecosocialist revolutionary change exist in all societies.

Such a process cannot begin without a revolutionary transformation of social and political structures based on the active support, by the majority of the population, of an ecosocialist program. The struggle of labor—workers, farmers, the landless and the unemployed—for social justice is inseparable from the struggle for environmental justice. Capitalism, socially and ecologically exploitative and polluting, is the enemy of nature and of labor alike.

Ecosocialism proposes radical transformations in:

- the energy system, by replacing carbon-based fuels and bio-fuels with clean sources of power under community control: wind, geothermal, wave, and above all, solar power.
- the transportation system, by drastically reducing the use of private trucks and cars, replacing them with free and efficient public transportation;
- present patterns of production, consumption, and building, which are based on waste, inbuilt obsolescence, competition, and pollution, by producing only sustainable and recyclable goods and developing green architecture;
- food production and distribution, by defending local food sovereignty as far as this is possible, eliminating polluting industrial agribusinesses, creating sustainable agro-ecosystems, and working actively to renew soil fertility.

To theorize and to work toward realizing the goal of green socialism does not mean that we should not also fight for concrete and urgent reforms right now. Without any illusions about "clean capitalism," we must work to impose on the powers-that-be—governments, corporations, international institutions—some elementary but essential immediate changes:

- drastic and enforceable reduction in the emission of greenhouse gases
- development of clean energy sources
- provision of an extensive free public transportation system
- progressive replacement of trucks by trains
- creation of pollution cleanup programs
- elimination of nuclear energy and war spending

These and similar demands are at the heart of the agenda of the Global Justice movement and the World Social Forums, which have promoted, since Seattle in 1999,[17] the convergence of social and environmental movements in a common struggle against the capitalist system.

Environmental devastation will not be stopped in conference rooms and treaty negotiations: only mass action can make a difference. Urban and rural workers, peoples of the Global South and indigenous peoples everywhere are at the forefront of this struggle against environmental and social injustice, fighting exploitative and polluting multinationals, poisonous and disenfranchising agribusinesses, invasive genetically modified seeds, and biofuels that only aggravate the current food crisis. We must further these social-environmental movements and build solidarity between anti-capitalist ecological mobilizations in the North and the South.

This Ecosocialist Declaration is a call to action. The entrenched ruling classes are powerful, yet the capitalist system reveals itself every day more financially and ideologically bankrupt, unable to overcome the economic, ecological, social, food and other crises it engenders. And the forces of radical opposition are alive and vital. On all levels, local, regional and international, we are fighting to create an alternative system based in social and ecological justice.

--------⚓--------

What about the USSR?

This book focuses on the connection between capitalism and the global ecological crisis, but it would be dishonest not to address the fact that some of the worst ecological nightmares of the twentieth century occurred in countries that called themselves socialist.

Karl Marx, exasperated by some of his French followers, once commented, "All I know is that I am not a Marxist." If he had lived through the twentieth century, he would probably have said the same thing about the environmental policies of regimes that claimed to be his political heirs.

To cite just one example, in the 1960s Soviet authorities launched a massive river-diversion project in Kazakhstan, Uzbekistan, and

Turkmenistan to irrigate new cotton plantations. The plantations bloomed and the USSR quickly became the second-largest cotton exporter in the world—but the region as a whole suffered unprecedented ecological disaster. The diverted rivers had fed the Aral Sea, then the fourth-largest lake in the world, comparable to Lake Huron. By 1989 it was less than 10 percent its original size. The remaining water was heavily polluted, groundwater was fouled, farms were destroyed by blowing salt, and a once vibrant fishing industry was gone.

We could add the nuclear horror of Chernobyl, or the fact that in the 1980s the USSR was the world's second-largest producer of greenhouse gas emissions. The Soviet Union had excellent environmental laws on paper, but air and water pollution were chronic problems.

People in the Soviet Union and the other Soviet Bloc countries thought they were building socialism. For most people worldwide that was what socialism looked like. So whether we call those societies socialist or give them some other label, they were clearly not solidarian societies, and were not on a path toward ecological civilization. What makes us think that the next attempts to build socialist societies will do any better?

In the 1920s and early 1930s, the Soviet Union was a world leader in ecological science and environmental protection. It was the first country to set aside large conservation areas, and one of the first to legally protect endangered species from hunting. There was strong support for scientists such as V. I. Vernadsky, who developed the theory of the biosphere, and N. I. Vavilov, who first traced the genetic origins of the world's major food plants.

Tragically, the political caste headed by Stalin abandoned the Marxist view of socialism as sustainable human development, arguing that the Soviet Union could outcompete capitalism through a forced march to full industrialization, without regard for human and environmental costs. Under Stalin's rule, the ecological movement was crushed, conservation areas were eliminated, and massive resources were thrown into the development of unconstrained heavy industry. Environmentalists who objected were imprisoned or executed.

Ecological science in the USSR was reborn in the late 1950s, and in many ways it surpassed the West. To cite just one example, Mikhail Ivanovich Budyko raised concerns about anthropogenic global warming in the 1960s, and his 1980 book *Global Ecology* advanced many of the concepts we now call Earth System science. After the collapse of the Soviet Union his work became better known in the rest of the world, and he was awarded the prestigious Blue Planet Prize in 1998.[18]

The Soviet state in the 1970s introduced environmental reforms in response to widespread environmental devastation and a powerful environmental movement led by scientists, but calls by prominent figures like Evgeni Federov for more rapid and radical changes went unheeded, with tragic results.

The destructive policies of Stalin and his successors were a world-historic catastrophe, but the experience also shows that an alternative path was possible. The adoption of ultra-productivist and anti-environmental policies was a defeat for the socialist cause in the USSR, not its result. As Oswaldo Martinez, president of the Economic Affairs Commission of Cuba's National Assembly, said in a 2009 interview, the USSR's experiences are a lesson for socialists in the twenty-first century.

> The socialism practiced by the countries of the Socialist Camp replicated the development model of capitalism, in the sense that socialism was conceived as a quantitative result of growth in productive forces. It thus established a purely quantitative competition with capitalism, and development consisted in achieving this without taking into account that the capitalist model of development is the structuring of a consumer society that is inconceivable for humanity as a whole.
>
> The planet would not survive. It is impossible to replicate the model of one car for each family, the model of the idyllic North American society, Hollywood etc.—absolutely impossible, and

this cannot be the reality for the 250 million inhabitants of the United States, with a huge rearguard of poverty in the rest of the world.

It is therefore necessary to come up with another model of development that is compatible with the environment and has a much more collective way of functioning.[19]

The environmental failures of the Soviet Bloc in the twentieth century demonstrate why ecology must have a central place in socialist theory, in the socialist program and in all socialist activity. There are no guarantees, but our only hope lies in building a movement that is deeply committed to replacing capitalism with an ecological civilization.

13

The Movement We Need

Only mass social movements can save us now. Because we know where the current system, left unchecked, is headed. We also know, I would add, how that system will deal with the reality of serial climate-related disasters: with profiteering, and escalating barbarism to segregate the losers from the winners. To arrive at that dystopia, all we need to do is keep barreling down the road we are on. The only remaining variable is whether some countervailing power will emerge to block the road, and simultaneously clear some alternate pathways to destinations that are safer. If that happens, well, it changes everything.

—NAOMI KLEIN[1]

If you don't know where you want to go, no road will take you there. However, knowing *where* you want to go is only the first part; it's not at all the same as knowing *how* to get there.

—MICHAEL LEBOWITZ[2]

The Holocene is over. The Anthropocene has begun.

That cannot be reversed. The climate changes already under way will last for thousands of years. No currently available technology can restore extinct species to their former abundance. The acid in oceans cannot be removed. Many glaciers have melted and much polar ice is gone forever. The oceans will continue rising.

Whether or not geologists formally decide to amend their official time scale, there is no doubt that the Earth System has entered a new epoch, one in which "human activities have become so pervasive and profound that they rival the great forces of nature and are pushing the Earth into planetary *terra incognita*."[3] In Barry Commoner's words, "the present system of production is self-destructive; the present course of human civilization is suicidal."[4]

The question is not whether the Earth System is changing, but how much it will change, and how we will live on a changed planet.

The question is not whether human activity can change Earth, but whether that power continues to be exercised for short-term private gain and destruction, or becomes a force for the long-term common good.

An ecological civilization will not "just happen." It can be made possible only by a deliberate and focused movement for change, a movement that works to achieve every change that is possible while capital still rules, and that consciously lays the groundwork for ousting capital in the future.

Ecological Counterpower

In 1864, in the manifesto that introduced the First International, Karl Marx described how the British workers' movement, not yet able to put an end to capitalism, had forced Parliament to enact laws that limited the bosses' power to exploit labor, by limiting the length of the working day. Marx described that campaign as part of "the great contest between the blind rule of the supply and demand laws which form the political economy of the middle class, and social production controlled by social foresight, which forms the political economy of the working class." The ensuing victory, he said, "was not only a great practical success; it was the victory of a principle; it was the first time that in broad daylight the political economy of the middle class succumbed to the political economy of the working class."[5]

Today, when we are not yet strong enough to win permanent solutions by ending the capitalist system, we must work to build counterpower that can force implementation of *ecological political economy* wherever possible. We may not be able to win lasting solutions now, but we can make the political and economic costs of inaction unacceptable to our capitalist rulers, and in doing so we can win time for Earth and for humanity.

Increasingly, the planetary emergency directly affects the daily lives of working people, farmers, indigenous communities, and all of the oppressed. As capitalism continues its relentless drive to expand no matter what damage it causes, we will see—we are already seeing—increasing resistance. Many of these struggles will focus on narrow issues, and many of the leaders and participants will have illusions about what can be done within the system. That's inevitable.

The worst mistake socialists can make in such circumstances— and unfortunately it's a mistake that many socialists do make—is to stand on the sidelines complaining that a given campaign isn't radical enough, or that it doesn't fit someone's preconceptions of what a movement ought to be.

Lenin famously warned against a narrow understanding of class struggle. He said socialists must be *tribunes of the people*, responding to "every manifestation of tyranny and oppression, no matter where it appears, no matter what stratum or class of the people."[6] In our time, socialists cannot be tribunes of the people unless we are also *tribunes of the environment*. We must respond, to the best of our ability, to every manifestation of capitalist environmental destruction.

We need to remember Marx's great insight that people in large numbers don't change themselves and then change the world, *they change themselves by changing the world*, and Rosa Luxemburg's argument that political education, class-consciousness, and organization "cannot be fulfilled by pamphlets and leaflets, but only by the living political school, by the fight and in the fight."[7]

Chilean Marxist Marta Harnecker puts it this way:

Being radical is not a matter of advancing the most radical slogans, or of carrying out the most radical actions, which only a few join in because they scare off most people. Being radical lies rather in creating spaces where broad sectors can come together and struggle. For as human beings we grow and transform ourselves in the struggle. Understanding that we are many and are fighting for the same objectives is what makes us strong and radicalizes us.[8]

A Majority Movement

Social and ecological changes as sweeping as those we need today will not happen just because they are the right thing to do. Good ideas are not enough. Moral authority isn't enough.

An ecosocialist revolution cannot be made by a minority. It cannot be imposed by politicians and bureaucrats, no matter how well-meaning they might be. It will require the active participation of the great majority of the people. In Marx's famous words: "The emancipation of the working classes must be conquered by the working classes themselves."[9]

This is not simply because democracy is morally superior, but because the necessary changes cannot be carried through, and will not be long-lasting unless they are actively supported, created, and implemented by the broadest possible range of people. Only majority support and commitment can possibly overcome the opponents of change.

The only way to overcome the forces that now rule, the forces of global destruction, is to organize a counterforce that can stop them and remove them from power. There is no such thing as a win-win revolution, where everyone gains and no one loses. In a real revolution, the people who had power and privileges in the old society lose them in the new.

We only have to look at the present U.S. Congress to see powerful people who will resist change even to the point of destroying the world. They are backed by some of the world's richest corporations, and they are prepared to bring the world down to protect their power.

And we have only to look at our own movements to recognize that ecosocialists—in fact, all kinds of socialists put together!—are a minority today. Not just in society at large, but also within the environmental movement. As Marxist scholar Fredric Jameson has written, we live in a time when most people find it easier to imagine the end of the world than to imagine the end of capitalism. Most green activists do not see capitalism as the primary problem—or, if they do, they don't believe an ecosocialist revolution is possible or desirable.

So the challenge for socialists is not to proclaim the revolution from every street corner, but rather to unite the broadest possible range of people, socialists or not, who agree that the climate vandals must be stopped. We need to work with everyone who is willing to join in fighting climate change in general, and the fossil fuel industry specifically.

Contrary to the pale greens who think we should place our faith in liberal politicians, and contrary to the advocates of guerilla attacks on infrastructure, there really is no shortcut to "creating spaces where broad sectors can come together and struggle."

A Tale of Two Cities

What forces will determine the outcome of the planetary crisis in the 21st century? A few years ago we had a strong foretaste of the class lineup.

In December 2009, the world's rich countries sent delegations to Copenhagen with instructions not to save the climate, but to block any action that might weaken their capitalist economies or harm their competitive positions in world markets. They succeeded. The backroom deal imposed by Obama was, as Fidel Castro wrote, "nothing more than a joke."

Five months after the Copenhagen meeting, in April 2010, a very different meeting took place in Cochabamba, Bolivia. At the invitation of Bolivian president Evo Morales, some 35,000 activists, many of them indigenous people, came from more than 130 countries to do what Obama and his allies refused to do in Copenhagen and since—develop a concrete program to save the Earth System.

They drafted a People's Agreement that places responsibility for the climate crisis on the capitalist system and on the rich countries that "have a carbon footprint five times larger than the planet can bear." They adopted eighteen major statements, covering topics from climate refugees to indigenous rights to technology transfer, and much more. It is impossible to imagine such a program coming out of any meeting of the wealthy powers, or any United Nations conference.

Those two meetings, in Copenhagen and Cochabamba, symbolize the great divide in the struggle for the future of the earth and humanity. On one side, the rich and powerful, determined to save their wealth and privileges, even if the world burns. On the other, indigenous people, small farmers and peasants, progressive activists and working people of all kinds, determined to save the world from the rich and powerful.

The Cochabamba conference showed, in a preliminary way, the alliance of forces that must be forged to end the environmentally destructive capitalist system. The movement needs students and academics and feminists and scientists, but to change the world we need the active participation of all oppressed people.

The Movement We Need

As we saw in chapter 9, Barry Commoner and other radical environmentalists realized that a new ecological regime had begun after the Second World War, long before it was identified and named by Earth System scientists. John Bellamy Foster described the postwar transformation as "a qualitative change in the level of human destructiveness."[10]

Recent scientific research has thoroughly confirmed that conclusion, but it is still a minority position on the left. Some prominent socialists deride such views as "catastrophism," and others still treat the environment as just one of many concerns and possibly a diversion from the "real" class struggle. Even among socialist environmentalists there has been little recognition of the qualitative changes to the entire Earth System that define the Anthropocene.

Building a movement that views the fight against capitalism and against global environmental destruction as inextricably linked poses unique challenges, and I don't pretend to have a blueprint. Indeed, one of the lessons we can learn from the failures of socialism in the twentieth century is that centrally dictated, one-size-fits-all plans for movement building will always fail. But in my view, for ecosocialist movements to have any chance of success in the Anthropocene, they must share four characteristics.

1. We must be pluralist and open to differing views within the green left.

Another lesson of the twentieth century is that monolithic socialist grouplets do not turn into mass movements. They stagnate and decay, they argue and they split, but they don't change the world. So no one should take my arguments as encouragement to found yet another leftist sect, or to reduce ecosocialism to what Marx derided as a *shibboleth*—a narrow schema that stands in the way of building a broad and united green-left current.[11] As Marta Harnecker argues, real unity is incompatible with either ignoring or suppressing differences.

> When we talk of unifying we are thinking of "grouping together," "uniting" the various actors around these goals which are of common Interest. Unify by no means implies "to make uniform," "to homogenize" nor does it mean to suppress differences but rather to act in common, building on the different characteristics of each group.[12]

Our ecosocialist programs define who we are: they are the glue that holds us together. But within that broad framework, we need to understand that none of us has a monopoly on truth and none of us has the magical keys to the ecosocialist kingdom. We will undoubtedly disagree on many issues, and our debates will be vigorous. But if we agree that there can be no true ecological revolution that is not socialist and no true socialist revolution that is not ecological, then what unites us is more important than our differences.

2. We must constantly extend our analysis and program in the light of changing political circumstances and scientific knowledge.

There is not, and there will never be, a perfect and immutable ecosocialist program, no document we can point to and say, "that's it, no more changes, we know what to do forever." We have beginning points; now we need to build on them, using the method of Marxism, the best scientific work of our time, and the real world experiences of struggles for change in a wide variety of places and situations.

In the past century, many Marxists treated Marxism—usually in some particular form, interpreted by one or another socialist leader—as a set of sacred texts that contained all the answers. That approach is alien to Marxism, which gives us a *method*, but not immutable answers. There is no substitute for concrete examination of the processes and trends that are driving social, political, economic, and environmental change: without such observation and analysis, ecosocialism will be irrelevant in the real world we need to change.

Equally alien to Marxism is the "postmodern" rejection of natural science that sometimes passes for profound radical thought. While postmodernism as such is now less trendy than it was, its continuing influence can be seen in the kneejerk hostility of some leftist academics toward the science of the Anthropocene and even toward the scientists themselves. Determined to build barriers against possible infection by new ideas, they entirely miss the significance of discoveries being made by people who, as Dipesh Chakrabarty points out, "are not necessarily anticapitalist scholars, and yet clearly . . . are not for business-as-usual capitalism either."[13]

We can learn from the example of Marx and Engels, who carefully studied the scientific and technological discoveries of their time, and worked to understand how new scientific knowledge could extend, deepen, or change their understanding of humanity's complex relationship with the rest of nature.

Dealing appropriately with changing circumstances and new scientific knowledge is difficult, because it requires us to think and respond creatively, not just repeat yesterday's slogans, but it's essential.

3. We must be internationalist and anti-imperialist.

All serious environmentalists must be internationalists, because ecosystems don't respect national borders. There is, in particular, no national solution to climate change. The necessary countermeasures must be fought for within each country, but only international change can defeat it. International communication, collaboration, and solidarity are absolutely essential.

Today, the most powerful and important struggles for environmental justice are taking place in the Third World. At the barest minimum, ecosocialists in the imperialist countries need to publicize those movements and build support for them. We need to show our solidarity as concretely as we can—but there must be much more to our internationalism than just cheerleading for remote struggles. We must be strong voices for concrete environmental justice.

It's been said many times that the people of the South, and indigenous people everywhere, are the primary victims of climate change and other forms of environmental destruction. What isn't said as often, but is even more important, is that the primary environmental criminals are "our" capitalists in the North. That places a special responsibility on ecosocialists in the wealthy countries to combat the policies of our governments and of the corporations that are based in our countries.

We must give particular emphasis and support to the demands raised in the Cochabamba People's Agreement, including:

- Financial support for adaptation to climate change, including the development of ecologically sound agriculture.
- Direct transfer of renewable energy and other technologies, so that the poorest countries can have economic development without contributing to global warming.
- Opposition to so-called market solutions, and to the commodification of nature. This includes rejecting carbon trading in all its forms.
- Welcoming climate refugees to our countries, offering them decent lives with full human rights.

4. We must actively participate in and build environmental struggles, large and small.

To be blunt, if we can't stop a pipeline or prevent fracking or get a university to stop investing in the oil industry, how can we imagine that we're actually going to overthrow capitalism? A socialist movement that doesn't take defending human survival as a central goal isn't worthy of the name.

We need to slow capitalism's ecocidal drive as much as possible and to reverse it where we can, to win every possible victory over the forces of destruction. As I've said, our rulers will not willingly change—but mass opposition can force them to act, even against their will. Our watchwords must be: Leave the oil in the soil, leave the coal in the hole, leave the tar sand in the land.

Our goal must be to bring together everyone—socialists, liberals, deep greens, trade unionists, feminists, indigenous activists and more—*everyone* who is willing to demand decisive action to bring down greenhouse gas emissions. To quote E. P. Thompson again, in the fight against exterminism, "secondary differences must be subordinated to the human ecological imperative":

> The immobilism sometimes found on the Marxist Left is founded on a great error: that theoretical rigor, or throwing oneself into a "revolutionary" posture, is the end of politics. The end of politics is to act, and to act *with effect*.[14]

Clearly, we also need to unite conscious ecosocialists, but the two tasks are not in conflict. Fighting for immediate gains against capitalist destruction and fighting for the ecosocialist future are not separate activities, they are aspects of one integrated process.

It is through united struggles for immediate gains and environmental reforms that working people and farmers and indigenous people can build the organizations and the collective knowledge they need to defend themselves and advance their interests. The victories they win in partial struggles help to build the confidence needed to take on bigger targets.

There are no guarantees. Marxism is not deterministic. An ecosocialist revolution is not inevitable. It will only happen if people consciously decide it is necessary, and take the steps needed to bring it about. Marx and Engels posed the alternative: the class struggle will lead either to "a revolutionary reconstitution of society at large" or to "the common ruin of the contending classes."[15]

In the Anthropocene, the common ruin of all, the destruction of civilization, is a very real possibility. That's why we need a movement with a clear vision, an ecosocialist program that can bridge the gap between the spontaneous anger of millions of people and the beginning of an ecosocialist transformation.

Marx famously wrote that humanity makes its own history, but not under conditions of its own choosing. The Anthropocene offers a powerful illustration of that truth, one that Marx could not have expected. We now face the challenge of changing the world in the context of impending environmental disaster on a global scale. That's reality in our time. The way we build socialism, the kind of socialism we will be *able* to build, will be fundamentally shaped by the state of the planet we must build it on. The longer it takes to get the necessary changes under way, the more difficult the transformation will be.

Antonio Gramsci's aphorism "Pessimism of the intellect, optimism of the will" defines our attitude in the Anthropocene. We know that disaster is possible, but we refuse to surrender to despair. If we fight, we *may* lose; if we don't fight, we *will* lose. Good or bad luck may play a role, but a conscious and collective struggle to stop capitalism's hell-bound train is our only hope for a better world.

As Gramsci also said: "It is necessary, with bold spirit and in good conscience, to save civilization. We must halt the dissolution which corrodes and corrupts the roots of human society. The bare and barren tree can be made green again. Are we not ready?"[16]

CONFUSIONS AND MISCONCEPTIONS

There is a history in green circles of blaming environmental problems on human beings as such. Our species has been labeled a plague, a virus, and a cancer, and compared to a swarm of locusts; we're told that people are nature's permanent enemy, that without radical population reduction all other environmental protection measures will certainly fail.

As Murray Bookchin wrote, neo-Malthusian greens blame environmental crises on "a vague species called humanity—as though people of color were equatable with whites, women with men, the Third World with the First, the poor with the rich, and the exploited with their exploiters."[1]

Given the prevalence of "blame people" views in conservative green circles, it not surprising that some radicals have reacted with suspicion to an epoch named for the *anthropos*, human beings. The following essays respond to two such concerns that have some currency on the left: the view that Anthropocene science blames all humanity for the planetary crisis, and the related assertion that scientists have chosen an inappropriate name for the new epoch.

1. Does Anthropocene Science Blame All Humanity?

It is clear that the world's poorest people are suffering most from climate change, and that their situation will get much worse if present trends continue. The injustice of that is especially appalling because, as study after study shows, the hardest hit are those who are least responsible. Stephen Pacala of Princeton University's

Environment Institute, for example, calculates that "the 3 billion poorest people . . . emit essentially nothing. . . . The development of the desperately poor is not in conflict with solving the climate problem, which is a problem of the very rich."[2]

That fact is so widely known and accepted that it is shocking to read the charge, made by some left-wing writers, that Anthropocene scientists blame people in general for global change—that "the Anthropocene narrative" views humanity as an undifferentiated whole and ignores differences between countries, classes, and institutions. For example:

- Keiran Suckling of the Center for Biological Diversity objects that the name identifies the cause of change as "humanity as a whole, rather than the identifiable power structures most responsible for the geological Anthropocene traces."[3]
- World-ecology theorist Jason Moore says that in the work of Anthropocene scientists, "the mosaic of human activity in the web of life is reduced to an abstract Humanity: a homogenous acting unit." He accuses them of treating "humanity as an undifferentiated whole" and offering "a meta-theory of humanity as collective agent."[4]
- Australian environmentalist Jeremy Baskin warns that "the Anthropocene label tends to universalize and normalize a small portion of humanity as 'the human of the Anthropocene.' . . . Impacts which have been driven by (and largely for the benefit of) a minority are attributed to all of humanity."[5]
- There's even been an Internet petition accusing geologists who support declaring a new Anthropocene epoch of "encouraging fatalism and myths about the wretchedness of human nature," and blaming environmental problems on "some essential 'human' quality."[6]

These would be serious charges, if they were true. It would mean that some of the world's most respected scientists are ignoring obvious facts. Worse, it would mean that those scientists are allied with the reactionary populationists who propose to save the world by letting billions of people die.

Fortunately, it isn't true. The criticisms reflect preconceptions about what the Anthropocene concept *might* mean, rather than serious engagement with the work of the scientists who have defined it.

That is not to say there are no people-are-the-problem advocates writing about the Anthropocene. Scientists are no more immune to mistaken social views than anyone else, and the word has been adopted by people from many fields—poets, philosophers, musicians, literary critics, journalists, and more—who use the Anthropocene as a hook to hang their particular preconceptions on.

The real surprise is how few neo-Malthusian passages there actually are in the scientific literature about the Anthropocene. Population growth is frequently mentioned as one of a number of factors associated with the Great Acceleration, but rarely is it identified as the main problem, nor is population reduction promoted as the *sine qua non* of any effective response to global change.

Indeed, overpopulationist ideologues are among the most hostile *opponents* of the Anthropocene project. A case in point is sociologist Eileen Crist, a prominent advocate of global population reduction: she has written pages of purple prose denouncing "Anthropocene discourse" as "a human species-supremacist planetary politics"; as an example of "the human supremacy complex"; as a "time-honored narrative of human ascent into a distinguished species" that "delivers a familiar anthropocentric credo" and "crystallizes human dominion . . . viewing our master identity as manifestly destined, quasi-natural, and sort of awesome." These are reasons, she writes, "to blockade the word Anthropocene" before it catches on.[7]

As Jedediah Purdy writes, "The Anthropocene does not seem to change many minds, strictly speaking, on point of their cherished convictions. But it does turn them up to 11."[8]

If the critics were challenging common misunderstandings about or misrepresentations of the Anthropocene, they would be on firmer ground, but they are not. They attack the entire field of Anthropocene study as inherently problematic, and attribute the people-are-the-problem view to specific scientists, including, by name, such leaders in the field as Paul Crutzen, Will Steffen, and Jan Zalaciewicz.

These critics are so convinced that natural scientists do not understand social issues that they fail to notice a substantial body of contrary evidence. If they actually read the scientific papers they cite in their footnotes, they must have been wearing ideological blinders.

For example, virtually every article on the Anthropocene mentions Paul Crutzen's 2002 article "The Geology of Mankind," which was the first paper on the subject published in a major journal. In it, Crutzen very clearly says that "these effects have largely been caused by only 25% of the world population."[9] One might question his statistics, or his social views in general, but it is obviously false to say that he treats humanity as an undifferentiated whole.

Crutzen's statement doesn't stand alone. The scientists in the forefront of the Anthropocene project have repeatedly and explicitly rejected any "all humans are to blame" narrative. The most authoritative book on the science of the Anthropocene, *Global Change and the Earth System,* includes passages such as these:

- "Present trends suggest that the gap between the wealthy and the poor is increasing almost universally, both within countries and between countries. . . . [Wealth differences] are often linked to different political economies and their effect on the ability of countries and locals to protect resources or enforce rules in their use. Wealth differences between countries have been shown to have significant impacts on resource use."[10]
- "An emphasis on the population variable can have the effect of blaming the victims (as in high fertility rates among economically marginal households in the tropical world) for consequences such as tropical deforestation and famine-malnutrition. In fact, modern famine and malnutrition are more

closely related to issues of food entitlements and endowments than to population growth."[11]

- "Population pressure and poverty have often been cited as the primary causes of tropical deforestation. However, a careful analysis of a large number of case studies across the tropics suggests that a more complex array of drivers including market and policy failures and terms of trade and debt are likely influences on the patterns and trajectories of land-use change in the tropics. As noted in one extensive review of the literature, forests fall because it is profitable to someone or some group."[12]

- "One-quarter of the world's population remains in severe poverty. Inequality has been increasing in many countries and between countries, and the interactions between poverty and the environment are of local, regional, and global significance."[13]

- "In a world in which the disparity between the wealthy and the poor, both within and between countries, is growing, equity issues are important in any consideration of global environmental management."[14]

A peer-reviewed article published in 2011, by some of the most prominent figures in Anthropocene science, is even clearer:

The post-2000 increase in growth rates of some non-OECD economies (e.g., China and India) is evident, but the OECD countries still accounted for about 75% of the world's economic activity. On the other hand, the non-OECD countries continue to dominate the trend in population growth. Comparing these two trends demonstrates that consumption in the OECD countries, rather than population growth in the rest of the world, has been the more important driver of change during the Great Acceleration.

The world's wealthy countries account for 80% of the cumulative emissions of CO_2 since 1751; cumulative emissions are important for climate given the long lifetime of CO_2 in the atmosphere. The world's poorest countries, with a combined population of about 800 million, have contributed less than 1% of the cumulative emissions.[15]

If those examples aren't enough to disprove the charge that Anthropocene science blames all of humanity for the actions of a small minority, we can turn to two landmark papers published in 2015: the update to the Planetary Boundaries Framework and the update to the Great Acceleration statistics and graphs, discussed in chapter 4. The authors of the former wrote:

> The current levels of the boundary processes, and the transgressions of boundaries that have already occurred, are unevenly caused by different human societies and different social groups. The wealth benefits that these transgressions have brought are also unevenly distributed socially and geographically. It is easy to foresee that uneven distribution of causation and benefits will continue, and these differentials must surely be addressed for a Holocene-like Earth System state to be successfully legitimated and maintained.[16]

And in their update on the Great Acceleration, scientists associated with the IGBP wrote:

> In 2010 the OECD countries accounted for 74% of global GDP but only 18% of the global population. Insofar as the imprint on the Earth System scales with consumption, most of the human imprint on the Earth System is coming from the OECD world. This points to the profound scale of global inequality, which distorts the distribution of the benefits of the Great Acceleration and confounds efforts to deal with its impacts on the Earth System....
>
> The Great Acceleration has, until very recently, been almost entirely driven by a small fraction of the human population, those in developed countries.[17]

I am not suggesting that the social analysis offered by Earth System scientists to date is complete, or even adequate: on the contrary, the problem of inequality is much more serious than even these passages suggest. Nevertheless, the charge that Anthropocene science as such blames all of humanity for the actions of a small minority and ignores inequalities of wealth and power simply doesn't hold water.

2. What's in a Name?

In one of Douglas Adams's *Hitchhiker's Guide to the Galaxy* books, a committee of marketing managers, stranded on a prehistoric planet, is unable to invent the wheel. Responding to a critic, the committee chair says, "Okay, if you're so clever, you tell us what color it should be!"

I'm reminded of that scene every time I read yet another article that responds to one of the most important scientific developments of our time, the Anthropocene, with the complaint that the scientists got the name wrong.

Never mind all that stuff about the Earth System changing in unprecedented and dangerous ways—*it needs a different name!*

The critics don't like the Greek root word *anthropos*, meaning human being—they fear it implies that every human on Earth is responsible for environmental destruction. Alternative suggestions include obvious jokes like Misanthropocene and Anthrobscene, and more serious proposals like Technocene, Sociocene, Homogenocene, Econocene, and Capitalocene.

So far as I can tell, none of these has been submitted to the Anthropocene Working Group, where they could be formally evaluated. But since the suggestions reveal misunderstandings about the word itself, and about the conventions used in naming geological epochs, a short discussion is in order.

To begin with, the Anthropocene is proposed as a new geological epoch, so its name should at least try to follow geology's naming conventions. The alternative proposals simply add a new word in front of the suffix *-cene*, apparently believing that it means epoch or age, which it does not.

The suffix *-cene* comes from the Greek *kainos* meaning "recent." It was introduced by the nineteenth-century geologist Charles Lyell, who distinguished between various layers of rock by determining the proportions of extinct and non-extinct fossils each contained. Thus Miocene is from *meios—few* of the fossils are recent. Pliocene is from *pleios—more* of the fossils are recent. Pleistocene is from *pleistos—most* of the fossils are recent.

After the Pleistocene, Lyell added an interval that he simply called Recent, but in 1885 the International Geological Congress changed that to Holocene, from the Greek *holos*, for strata in which the fossils are *wholly or entirely* recent.

So, contrary to what is often said in magazine articles, Anthropocene does not mean Human Age or Human Era. It combines *kainos* with *anthropos*, meaning human being; so, following Lyell's approach, it means a time when geological strata are dominated by remains of recent human origin. Indeed, a key part of the ongoing Anthropocene debate among geologists concerns which such remains should be used to identify the new epoch. From the perspective of historical and physical geology, the name is appropriate.

In left-wing circles, the most often proposed alternative name for the new epoch is *Capitalocene*. Proponents argue that global change is being driven by a specific form of society, not humans in general, so the new epoch should be named after capitalism.

Most people who make that suggestion simply want to focus attention on capitalism's responsibility for the crisis in the Earth System. Although I don't think insisting on a name change is appropriate, I fully sympathize with the motivation: I think this book makes that very clear.

But a few academics go overboard, proposing that we accept *capitalism* and *capitalocene* as different names for the same thing: a new social/economic/environmental epoch that emerged in the 1500s.

Philosophers might call this a category mistake—capitalism is a 600-year-old social and economic system, while the Anthropocene is a 60-year-old Earth System epoch. Any serious engagement with social and natural science will conclude that capitalism existed for hundreds of years before the new geological epoch began, and that the new epoch will continue long after capitalism is a distant memory. Treating them as identical can only weaken efforts to get rid of capitalism and mitigate the harm it has caused to the Earth System, so that human society can survive—and hopefully prosper—in the Anthropocene.

(In passing: If our current epoch is the *Capitalocene*, then surely the previous epoch should be renamed *Feudalocene*, preceded by the *Slaveryocene*, preceded by . . . what? The *Hunter-Gathererocene*? The fact that no one suggests such absurdities is instructive.)

The root word *anthropos* also appears in another common Earth Science term, *anthropogenic*. The expression "anthropogenic climate change" does not mean that all humans cause global warming; rather, it distinguishes changes that are caused by human action from those that would have occurred whether or not humans were involved. Similarly, Anthropocene does not refer to all humans, but to an epoch of global change that would not have occurred in the absence of human activity.

So take a deep breath, folks. The *fact* of the Anthropocene raises important political issues, but there is no hidden political agenda in the *word*. Anthropocene does not imply a judgment about all humans or human nature.

The name is not perfect. As the often overheated discussions show, it is open to misinterpretation. Maybe if ecosocialists had been present when Paul Crutzen invented the word in 2000 a different name would have been adopted, but now Anthropocene is widely used by scientists and non-scientists alike. Insisting on a different word (for left-wing use only?) can only cause confusion, and direct attention away from far more important issues.

Let's focus on the wheel, and not get hung up on what color it ought to be.

NOTES

Foreword

1. Brecht, *Brecht on Theatre*, 275.
2. Hamilton and Grinevald, "Was the Anthropocene Anticipated?," 67.
3. Crutzen and Stoermer, "The Anthropocene"; Crutzen, "Geology of Mankind"; Waters et al., "The Anthropocene Is Functionally and Stratigraphically Distinct from the Holocene."
4. Weart, "Interview with M.I. Budyko"; Budyko, "Polar Ice and Climate"; Sellars, "A Global Climatic Model Based on the Energy Balance of the Earth-Atmosphere System"; Budyko, "Comment," 310.
5. Mészáros, *The Power of Ideology*, 128.
6. Marx and Engels, *MECW*, vol. 5, 40.
7. Marsh, *Man and Nature*; Golley, *A History of the Ecosystem Concept in Ecology*, 2, 207; Marx, *Capital*, vol. 1, 636–39, vol. 3, 949.
8. Margulis and Sagan, *What Is Life?*, 47; Vladimir I. Vernadsky, *The Biosphere* (New York: Springer-Verlag, 1998). The concept of the biosphere was originally introduced by the French geologist Edward Suess in 1875, but was developed much further by Vernadsky and hence came to be associated primarily with the latter.
9. Vernadsky in Ross, *150 Years of Vernadsky*, vol. 2, 82; Shantser, "The Anthropogenic System (Period)," 140. The word *Anthropocene* first appeared in English in 1973 in the Shantser article in *The Great Soviet Encyclopedia*.
10. Levins and Lewontin, *The Dialectical Biologist*, 277; Oparin, "The Origin of Life"; and Haldane, "The Origin of Life."
11. Carson, *Lost Woods*, 230–31.
12. See Hutchinson, "The Biosphere."
13. Commoner, *The Closing Circle*, 45–62, 138–75, 280.
14. Fedorov quoted in Brodine, *Green Shoots, Red Roots*, 29. See also Fedorov, *Man and Nature*, 29–30; Foster, "Late Soviet Ecology and the Planetary Crisis," 9; Budyko, *The Evolution of the Biosphere*, 406. The serious warnings of prominent figures like Fedorov calling for a more rapid and radical response by the Soviet state to environmental problems went largely unheeded, with tragic results.
15. Fedorov, *Man and Nature*, 146.

16.   Hamilton and Grinevald, "Was the Anthropocene Anticipated?," 64.
17.   Odum, *Environment, Power, and Society for the Twenty-First Century*, 3.
18.   Ibid., 263.
19.   Thompson, *Beyond the Cold War*, 41–80; Bahro, *Avoiding Social and Ecological Disaster*, 19; Odum, *Environment, Power, and Society for the Twenty-First Century*, 276–78.
20.   Edburg and Yablokov, *Tomorrow Will Be Too Late*.
21.   Brecht, *Tales from the Calendar*, 31–32.

Preface

1.   Commoner, *The Closing Circle*, 178.
2   Ibid., 277.

Part One: A No-Analog State

1.   In the final panel, Bob says, "Wow. That carbon dates me."
2.   Hamilton, "The Anthropocene: Too Serious for Postmodern Games."

1. A Second Copernican Revolution

1.   Amsterdam Declaration on Global Change, July 2001.
2.   Steffen, "Commentary," 486.
3.   Crutzen, "My Life with O_3, NO_x and Other YZO$_x$s."
4.   Quoted in Pearce, *With Speed and Violence*, 21.
5.   See http://highlycited.com.
6.   Steffen et al., "The Anthropocene: Are Humans Now Overwhelming the Great Forces of Nature?," 614.
7.   Steffen et al., *Global Change and the Earth System*, 1.
8.   Roederer, "ICSU Gives Green Light to IGBP."
9.   Ibid.
10.   "ICSU Sets Up IGBP Committee."
11.   Oldfield et al., "The Earth System," 7.
12.   Schellnhuber, "Earth System Analysis and the Second Copernican Revolution," 19–23.
13.   Steffen et al., *Global Change and the Earth System*, 264.
14.   http://www.igbp.net/publications/igbpbookseries.4.d8b4c3c12bf3be6 38a8000924.html.
15.   Steffen et al., "Trajectory of the Anthropocene," 82.
16.   Steffen, "Commentary," 486–87.
17.   Crutzen and Stoermer, "The Anthropocene," 17. Stoermer took no further part in Anthropocene discussions.
18.   Falkowski et al., "The Global Carbon Cycle," 291–96. Given the normal lead time for peer-reviewed articles, this article must have been completed shortly after the Crutzen-Stoermer article appeared. As will be discussed in chapter 3, if a geologist had written this, the word *era* would have been *epoch*.
19.   Steffen et al., *IGBP Science* 4, 11–12.

20. Will Steffen confirms that in 2001 the Anthropocene was "largely unknown" to the other sponsoring groups. Email to author, July 16, 2015.
21. Crutzen, "Geology of Mankind."
22. Steffen, email to author, July 16, 2015.
23. *Global Change and the Earth System* is now out of print, but it is available as a pdf on the IGBP website, http://www.igbp.net/.

2. The Great Acceleration

1. Commoner, *The Closing Circle*, 140. Emphasis in original.
2. Steffen et al., "The Trajectory of the Anthropocene," 82.
3. Ibid.
4. Steffen et al., *Global Change and the Earth System*, 131.
5. Millennium Ecosystem Assessment, news release, June 5, 2001.
6. Millennium Ecosystem Assessment, *Living Beyond Our Means,* 9.
7. Millennium Ecosystem Assessment, *Ecosystems and Human Well-Being,* 2–4. All the MEA reports are available at http://www.millenniumassessment.org/.
8. Steffen et al., *Global Change and the Earth System,* 131.
9. Hibbard et al., "Group Report," 342.
10. Steffen et al., "The Trajectory of the Anthropocene," 82.
11. Steffen et al., "The Anthropocene: Are Humans Now Overwhelming the Great Forces of Nature?," 617, 618.
12. Figures 2.1 and 2.2 graphs created by R. Jamil Jonna based on data in Steffen et al., "Trajectory of the Anthropocene," 81–98.
13. Steffen et al., "The Trajectory of the Anthropocene," 89.
14. Ibid., 91.

3. When Did the Anthropocene Begin?

1. Oreskes, "The Scientific Consensus," 93.
2. Zalasiewicz et al., "Are We Now Living in the Anthropocene?" 7.
3. See, for example, Syvitski et al., "Impact of Humans on the Flux of Terrestrial Sediment to the Global Coastal Ocean."
4. Zalasiewicz et al., "Are We Now Living in the Anthropocene?" The following quotes are from this article.
5. The requirements for formally defining a new geological interval, and issues that are specific to this case, are discussed in several essays in Waters et al., *A Stratigraphical Basis for the Anthropocene.*
6. Barnosky, "Did the Anthropocene Begin with a Bang or a Drumroll?"
7. Crutzen, "Geology of Mankind," 17.
8. Ruddiman, "The Anthropogenic Greenhouse Era Began Thousands of Years Ago"; Ruddiman, "How Did Humans First Alter Global Climate?"
9. Hamilton, "Can Humans Survive the Anthropocene?"
10. Hansen et al., "Ice Melt, Sea Level Rise, and Superstorms."
11. Nordhaus et al., "Ecomodernism and the Anthropocene." For a discussion

of the reactionary role of the Breakthrough Institute in Anthropocene discussions, see Angus, "Hijacking the Anthropocene."

12. Hamilton et al., "Was the Anthropocene Anticipated?"
13. Zalasiewicz et al., "When Did the Anthropocene Begin?" 6.
14. Ibid.
15. Vaughan, "Human Impact Has Pushed Earth into the Anthropocene, Scientists Say."

4. Tipping Points, Climate Chaos, and Planetary Boundaries

1. Rockström et al., "Planetary Boundaries."
2. Steffen et al., *Global Change and the Earth System*, 6, 7.
3. Alverson et al., "The Past Global Changes (PAGES) Program," 169–73.
4. Steffen et al., *Global Change and the Earth System*, 23.
5. Ibid., 14.
6. Ibid., 12.
7. Moore, "Challenges of a Changing Earth," 9.
8. Steffen, "An Integrated Approach to Understanding Earth's Metabolism," 9.
9. Useful accounts of this climate event include Kunzig, "Hothouse Earth"; Kump, "The Last Great Global Warming"; and Masters, "PETM: Global Warming, Naturally."
10. Moore, "Challenges of a Changing Earth," 9.
11. Weart, *The Discovery of Global Warming*.
12. Adams et al., "Sudden Climate Transitions during the Quaternary," 2.
13. National Research Council, *Abrupt Impacts of Climate Change: Anticipating Surprises*, 8.
14. Steffen et al., "Abrupt Changes," 9.
15. Marsh, *Man and Nature*, 34.
16. Leopold, *A Sand County Almanac*, 262.
17. Steffen et al., *Global Change and the Earth System*, 83.
18. Ganopolski et al., "Rapid Changes of Glacial Climate," 153–58.
19. Burroughs, *Climate Change in Prehistory*, 13, 102.
20. Zalasiewicz et al., *The Goldilocks Planet*, 204.
21. Ibid., 205. Emphasis in original.
22. Petit et al., "Climate and Atmospheric History," 429–36.
23. Barnosky et al., "Approaching a State Shift."
24. Alley et al., *Abrupt Climate Change*, v.
25. Ibid., 1.
26. Rockström et al., *Big World Small Planet*, 59.
27. Ibid., 60.
28. Rockström et al., "A Safe Operating Space for Humanity"; Rockström et al., "Planetary Boundaries: Exploring the Safe Operating Space for Humanity."
29. Steffen et al., "Planetary Boundaries: Guiding Human Development."
30. Ibid.
31. Rockström, "Bounding the Planetary Future."
32. Rockström et al., "A Safe Operating Space for Humanity," 474.

33. Steffen et al., *Global Change and the Earth System*, 235.
34. Broecker, "Cooling for the Tropics," 212–13.

5. The First Near-Catastrophe

1. Sweezy, "Capitalism and the Environment," 5.
2. McNeill, *Something New Under the Sun*, 111n, 113.
3. Cowan, "The Industrial Revolution in the Home," 181–201.
4. Marx, *Capital*, vol. 1, 617.
5. CFCs were also widely used as fire suppressants, under the brand name Halon, as well as in polyurethane foam insulation and packaging materials.
6. McNeill, *Something New Under the Sun*, 113.
7. Lovelock et al., "Halogenated Hydrocarbons," 194–96.
8. Cagin et al., *Between Earth and Sky*, 8.
9. Quoted in Roan, *Ozone Crisis*, 96.
10. Rowland, Nobel Lecture in Chemistry.
11. A similar process occurs in the North, but it is less severe because the Arctic Vortex winds are milder than the Antarctic Vortex.
12. For a comprehensive account of the adoption and implementation of the Montreal Protocol, see Andersen et al., *Protecting the Ozone Layer*.
13. McNeill, *Something New Under the Sun*, 112.
14. National Research Council, *Abrupt Impacts*, 36.
15. Crutzen, "My Life with O_3, NO_x and Other YZO_xs."
16. Ibid.
17. Maxwell et al., "There's Money in the Air," 281.
18. Quoted in Roan. *Ozone Crisis*. 144.

6. A New (and Deadly) Climate Regime

1. National Research Council, *Climate Stabilization Targets*, 19.
2. UNFCCC, Paris Agreement, December 12, 2015, Art. 2, Par. 1(a).
3. Quoted in *The Guardian*, December 12, 2015.
4. Climate Action Tracker, "2.7°C Is Not Enough—We Can Get Lower."
5. Hansen et al., "Perception of Climate Change."
6. Ibid.
7. Ibid.
8. Karl and Katz, "A New Face for Climate Dice." 14271.
9. Hansen et al., "Perception of Climate Change."
10. Coumou et al., "Decade of Weather Extremes," 496.
11. Seneviratne et al., "No Pause," 163.
12. Williams et al., "Contribution of Anthropogenic Warming": Lewis and King, "Dramatically Increased Rate."
13. Field et al., *Managing the Risks of Extreme Events*, 112.
14. Anderson et al., "Beyond 'Dangerous' Climate Change," 23.
15. Schellnhuber, "The Laws of Nature—and the Laws of Civilization."
16. Representative Concentration Pathways were introduced in 2012 to replace the emission scenarios that the IPCC had used since 2000. The RCPs are

essentially massive databases of climate-related data for the entire world, subdivided into more than 500,000 geographic cells. This permits highly detailed studies of global change from the late 1900s to about 2100. For a non-technical overview, see Wayne, *The Beginner's Guide to Representative Concentration Pathways.*

17. Potsdam Institute, *Turn Down the Heat* (2014), 6.
18. Potsdam Institute, *Turn Down the Heat* (2012), xv.
19. Potsdam Institute, *Turn Down the Heat* (2014), 5.
20. IPCC, "Summary for Policymakers," 8.
21. Diffenbaugh et al., "Observational and Model Evidence." The study used models based on the IPCC's A1B scenario, in which fossil intensive and non-fossil energy use expand at equal rates.
22. Ibid., 616.
23. *Lancet*, "Managing the Health Effects of Climate Change"; *Lancet*, "Health and Climate Change."
24. World Health Organization, *Climate Change and Health.*
25. Wet bulb globe temperature is a heat-stress index that combines temperature, humidity, wind, and solar radiation.
26. *Lancet*, "Climate Change and Health," 187.
27. Dunne et al., "Reductions in Labour Capacity," 565.
28. Potsdam Institute, *Turn Down the Heat* (2014), 16.
29. UNFCCC Secretariat, *Report on the Structured Expert Dialogue*, 16, 17.
30. Ibid. Emphasis added.
31. Archer, *The Long Thaw*, 1.
32. Marcott et al., "A Reconstruction of Regional and Global Temperature," 1201.
33. Risbey, "The New Climate Discourse: Alarmist or Alarming," 33.
34. Wagner et al., *Climate Shock*, 53.
35. Anderson et al., "A 2°C Target?" 19.
36. Klein, *This Changes Everything*, 88.

Part Two: Fossil Capitalism

1. Marx, "Speech at the Anniversary of the People's Paper," *MECW*, vol. 14, 655.
2. Marx, "The Future Results of British Rule in India," *MECW*, vol. 12, 222.
3. Steffen et al., "The Trajectory of the Anthropocene," 92–93. For clarity, I have paraphrased these points.
4. Ibid., 93–94.
5. Marx, *Capital*, vol. 3, 949.
6. Steffen et al., "The Anthropocene: From Global Change to Planetary Stewardship," 739.
7. Lenin, "Imperialism, the Highest Stage of Capitalism," 267.

7. Capital's Time vs. Nature's Time

1. Carson, *Lost Woods*, 194–95.

2.   White House, "The G-20 Toronto Summit Declaration," paragraph 7.
3.   Mill, *Essays on Some Unsettled Questions*, 116.
4.   Nadeau, *The Environmental Endgame*, 146, 166,
5.   Ibrahim, *Capitalism versus Planet Earth,* 164.
6.   Ibid., 161.
7.   *MECW*, vol. 5, 487.
8.   Harvey, *Seventeen Contradictions*, 232.
9.   Foster, *Marx's Ecology*, 160. Emphasis in original.
10. Tansley, "Use and Abuse of Vegetational Terms," 284.
11. Sweezy, "Capitalism and the Environment," 7–8.
12. *MECW*, vol. 42, 227.
13. Marx, *Capital*, vol. 1, 637.
14. Ibid., vol. 3, 195.
15. Ibid., 949.
16. Foster, *Marx's Ecology*, 163.
17. Weston, *Political Economy of Global Warming*, 65–66.
18. Carson, *Silent Spring*, 7.
19. Hamilton, "Human Destiny in the Anthropocene," 35.
20. Marx, *Capital,* vol. 3, 754n.
21. Ibid., vol. 2, 321–22.
22. Wright et al., *Climate Change, Capitalism and Corporations*, 45.
23. Mészáros, *Challenge and Burden*, 386, 383.
24. Marx, *Capital*, vol. 1, 860.
25. Commoner, "Threats to the Integrity of the Nitrogen Cycle," 70.
26. Weston, *Political Economy of Global Warming*, 67.

8. The Making of Fossil Capitalism

1.   Mészáros, *Necessity of Social Control*, 49–50.
2.   Engels to C. Schmidt, August 5, 1890, *MECW*, vol. 49, 8.
3.   Mandel, *Second Slump*, 28.
4.   *MECW,* vol. 46, 411.
5.   Marx, *Capital,* vol. 1, 1021.
6.   This account is obviously much oversimplified. For a full account of the transition from water to steam, and thus of the birth of fossil capitalism, I highly recommend Malm's book, *Fossil Capital*.
7.   See ibid., esp. chaps. 1–3.
8.   Foner, *Great Labor Uprising,* 13–14.
9.   Malm, *Fossil Capital*, 13, 253.
10. Simms, *Ecological Debt*, 97.
11. Podobnik, *Global Energy Shifts*, 29.
12. Ibid., 30–31.
13. Ibid., 27, 65.
14. Mitchell, *Carbon Democracy,* 63.
15. Pobodnik, *Global Energy Shifts*, 69.
16. Baran et al., *Monopoly Capital*, 219.

17.   DuBoff, *Accumulation and Power*, 83–84.
18.   Beckman, "Brief History," 159.
19.   Braverman, *Labor and Monopoly Capital*, 113.
20.   Mandel, *Marxist Economic Theory*, vol. 2, 431n.
21.   Mandel, *Late Capitalism*, 215.
22.   Braverman, *Labor and Monopoly Capital*, 108.
23.   Ross et al., *The Polluters*, 25.
24.   Chandler et al., *Scale and Scope*, 21.
25.   Ibid., 196–97.
26.   Mandel, *Marxist Economic Theory*, 398.
27.   Baran et al., *Monopoly Capital*, 6.
28.   Galbraith, *New Industrial State*, 92.
29.   Malm, *Fossil Capital*, 11.
30.   Commoner, *The Closing Circle*, 140.

9. War, Class Struggle, and Cheap Oil

1.   Suvin, *In Leviathan's Belly*, Kindle ed., loc. 1825.
2.   Hobsbawm, *Age of Extremes*, 258.
3.   Quoted in Kennedy, *Freedom from Fear*, 622.
4.   J. Thorne, "Profiteering in the Second World War." J. Thorne may have been Frank Lovell or Art Preis.
5.   Lipsitz, *Rainbow at Midnight*, 57; Heartfield, *Unpatriotic History*, 36.
6.   Lipsitz, *Rainbow at Midnight*, 61.
7.   Quoted in Preis, *Labor's Giant Step*, 301.
8.   Klare, *Blood and Oil*, 28.
9.   Mitchell, *Carbon Democracy*, 111; Klare, *Blood and Oil*, 28.
10.   Yergin, *The Prize*, 383–84.
11.   Spitz, *Petrochemicals*, 153–54, 228.
12.   Phillips, *Wealth and Democracy*, 77.
13.   Oakes, "Toward a Permanent War Economy?"
14.   Quoted in Du Boff, *Accumulation and Power*, 99.
15.   Jansson, *Sixteen-Trillion-Dollar Mistake*, 76.
16.   Ibid., 109.
17.   Preis, *Labor's Giant Step*, 378.
18.   Turgeon, *Bastard Keynesianism*, 10.
19.   Lipsitz, *Rainbow at Midnight*, 99.
20.   Preis, *Labor's Giant Step*, 276.
21.   Drucker, "What to Do About Strikes," 12.
22.   Cochran, *Labor and Communism*, 322.
23.   Yates, *Naming the System*, 220.
24.   Judt, *Postwar*, 91.
25.   *Chicago Tribune*, "Rockefeller Profits from the Marshall Plan," December 13, 1948.
26.   Mitchell, *Carbon Democracy*, 30.
27.   Machado, *In Search of a Usable Past*, 122–23.

28. Painter, "The Marshall Plan and Oil," 168.
29. Hobsbawm, *Age of Extremes*, 262.
30. Yergin, *The Prize*, 541–42.
31. Pobodnik, *Global Energy Shifts*, 92.
32. Carson, *Silent Spring*, 15, 16.
33. Bookchin, *Our Synthetic Environment*, 53.
34. Commoner, *The Closing Circle*, 144.
35. Commoner, *Making Peace with the Planet*, 7.
36. Foster, *Vulnerable Planet*, 109, 114.

10. Accelerating the Anthropocene
1. Weston, *The Political Economy of Global Warming*, 195.
2. Fressoz, "Losing the Earth Knowingly," 72.
3. Harman, *A People's History of the World*, 548–49.
4. Davis, *Prisoners of the American Dream*, 191.
5. Harrington, *The Other America*, 4.
6. Hobsbawm, *Age of Extremes*, 259.
7. Steffen et al., "The Trajectory of the Anthropocene," 91.
8. Quoted in Huber, *Lifeblood*, 42.
9. Yergin, *The Prize*, 409.
10. DuBoff, *Accumulation and Power,* 102.
11. Cohen, *Consumers Republic*, 199.
12. Harvey, *Enigma of Capital*, 106–7.
13. Brown, "Pavement Is Replacing the World's Croplands."
14. Kunstler, *The Long Emergency,* 248.
15. Lewontin, "Agricultural Research," 12; Levins, "Why Programs Fail," 45.
16. Dimitri et al., "20th-Century Transformation," 6.
17. Hoppe, *Structure and Finances of U.S. Farms*, iii.
18. Perfecto et al., *Nature's Matrix*, 50–51.
19. Huber, *Lifeblood*, 87.
20. Nikiforuk, *Energy of Slaves*, 85.
21. Ibid., 80.
22. Huber, *Lifeblood*, 87.
23. Clapp, *Food*, 52.
24. Foster et al., "The U.S. Imperial Triangle and Military Spending," 1.
25. Buxton et al., "Security for Whom in a Time of Climate Crisis?," 13.
26. Semova et al., "US Department of Defense Is the Worst Polluter on the Planet."
27. Smil, *Energy at the Crossroads*, 81.
28. Military emissions are also excluded under the 2015 Paris Agreement.
29. Sanders, *The Green Zone*, 51–55, 68.
30. Shah, *Crude*, 144–45.
31. Custers, *Questioning Globalized Militarism*, 12.
32. Hynes, "Pentagon Pollution, Part One."
33. Sanders, *The Green Zone*, 110.

34. Foster et al., *Endless Crisis*, 41.
35. Bivens et al., *Raising America's Pay*, 10.
36. Smith, "Outsourcing, Financialization and the Crisis," 20, 33.
37. Foster et al., *The Endless Crisis*, 103–23.
38. Ibid., 170.
39. Dowd, *Capitalism and Its Economics*, 182.
40. Malm, "China as Chimney of the World," 149.
41. Malm, *Fossil Capital*, 346.
42. Vidal, "Shipping Boom Fuels Rising Tide of Global CO2 Emissions."
43. Harrould-Kolieb, *Shipping Impacts on Climate*, 2.
44. Winebrake et al., "Mitigating the Health Impacts of Pollution from Oceangoing Shipping," 4776.
45. Eyring et al., "Emissions from International Shipping," 16.
46. Biel, *Entropy of Capitalism*, 124.
47. World Economic Forum. *The New Plastics Economy*, 6–15.
48. *GRAIN*, "The Exxons of Agriculture," 7.
49. Di Muzio, "Capitalizing a Future Unsustainable," 376–79.
50. United Nations, *World Economic and Social Survey 2011*, 90.
51. Coady et al., *How Large Are Global Energy Subsidies?*, 29, 31.
52. Malm, *Fossil Capital*, 288.
53. Ibid., 353. Emphasis in original.
54. Jacobson et al., "Providing All Global Energy with Wind, Water, and Solar Power, Part I"; Delucchi et al., "Providing All Global Energy with Wind, Water, and Solar Power, Part II"; Jacobson, Delucchi et al., "100% Clean and Renewable Wind, Water, and Sunlight."
55. Quoted in Klein, *This Changes Everything*, 102.
56. Harman, *Zombie Capitalism*, 311.

11. We Are Not All in This Together

1.   Roy, *An Ordinary Person's Guide to Empire*, 20–21.
2.   Quoted in Meyer, "Al Gore Dreamed Up a Satellite."
3.   Fuentes-Nievaet et al., "Working for the Few," 2.
4.   Global Humanitarian Forum, *Human Impact Report*, 60, 62.
5.   Klinenberg, *Heat Wave*, xxiii.
6.   Carty et al., *Growing Disruption*, 3.
7.   UNDP, *Human Development Report 2007/2008*, 89.
8.   Ruder, "How Katrina Unleashed a Storm of Racism," 8–9.
9.   Hartman and Squires, *There Is No Such Thing*, 4, 6.
10. E. P. Thompson, "Notes On Exterminism," 22.
11. Goff, "Exterminism and the World in the Wake of Katrina."
12. Hodges, "Drown an Immigrant to Save an Immigrant."
13. Kumar, "Green Climate Fund Faces Slew of Criticism," 419–420.
14. Parenti, "The Catastrophic Convergence," 33.
15. Kaplan, *The Coming Anarchy*, 19–20.
16. White House, *A National Security Strategy of Engagement and Enlargement*, 1.

17. Schwartz et al., *An Abrupt Climate Change Scenario and Its Implications for United States National Security*, 2, 18.
18. Quoted in Angus et al., *Too Many People?*, 111.
19. Commoner, *The Closing Circle*, 297.
20. Parenti, "The Catastrophic Convergence," 35.
21. Eckersley, "Environmental Security, Climate Change, and Globalizing Terrorism," 87.
22. Engels, "The Condition of the Working Class in England," *MECW*, vol. 4, 348.
23. UNDP, *Human Development Report 2007/2008*, 166.
24. Davis, "Who Will Build the Ark?," 37, 38.
25. Davis et al., *Evil Paradises*, Introduction.
26. Klein, *This Changes Everything*, 51.
27. Marx, *Capital*, vol. 1, 782, 929.
28. Harvey, *Seventeen Contradictions*, 249.
29. Mészáros, *Socialism or Barbarism*, 80.

Part Three: The Alternative
1. Speth, *The Bridge at the Edge of the World*, 1.
2. Linden, "Storm Warnings Ahead," 58.
3. Speth, *The Bridge at the Edge of the World*, 7.
4. Hansen, *Storms of My Grandchildren*, ix.
5. UNDP, *Human Development Report, 2007–2008*, 7.
6. Hamilton et al., "Thinking the Anthropocene," 5.

12. Ecosocialism and Human Solidarity
1. Marx, *Capital*, vol. 3, 911.
2. Singer, *Whose Millennium?*, 259.
3. Solnit, *A Paradise Built in Hell*, 1.
4. Ibid., 8, 2, 7, 21.
5. Lebowitz, *The Socialist Alternative*, 144. Emphasis in original.
6. Costa, "Socialism Is Not Possible on a Ruined Planet."
7. Singer, *Whose Millennium?*, 259.
8. Mészáros, *Challenge and Burden*, 250.
9. Magdoff, "Ecological Civilization," 20.
10. Morris et al., *Socialism: Its Growth & Outcome*, 285.
11. Mandel, *Long Waves*, 80–83. Paragraph breaks added for readability.
12. Angus and Butler, *Too Many People?*, 198–99.
13. Lebowitz, *The Socialist Alternative*, 131. Lebowitz describes this as a *partial* charter, because it is an alternative to capitalist relations, and "does not address other inversions of human development such as patriarchy, caste society and racism except implicitly."
14. Abramsky, "Racing to 'Save' the Economy and the Planet," 26.
15. The full text is published in Angus, *The Global Fight for Climate Justice*, 233–38.

16. Townsend, "Change the System—Not the Climate!" Edited and updated with the author's permission.
17. This refers to the mass protests that challenged the World Trade Organization meeting in Seattle in 1999.
18. For an account of Soviet ecological science after the 1950s, see Foster, "Late Soviet Ecology and the Planetary Crisis."
19. Martinez, "We Are Facing Something More than a Mere Financial Crisis."

13. The Movement We Need

1. Klein, *This Changes Everything*, 450.
2. Lebowitz, *The Socialist Alternative*, 127.
3. Steffen et al., "The Anthropocene: Are Humans Now Overwhelming the Great Forces of Nature?," 614.
4. Commoner, *The Closing Circle*, 295.
5. Marx, "Inaugural Address of the International Working Men's Association," *MECW*, vol. 20, 11.
6. Lenin, "What Is to Be Done," *Collected Works*, vol. 20, 423.
7. Luxemburg, "The Mass Strike, the Political Party, and the Trade Unions," *Rosa Luxemburg Reader*, 182.
8. Harnecker, *Rebuilding the Left*, 4.
9. Marx, "Provisional Rules of the Association," *MECW*, vol. 20, 14.
10. Foster, *Vulnerable Planet*, 114.
11. Marx: "The sect seeks its *raison d'etre* and its *point d'honneur*, not in what it has in common with the class movement, but in the *particular shibboleth distinguishing* it from that movement." *MECW*, vol. 43, 133.
12. Harnecker, *Rebuilding the Left*, 78.
13. Chakrabarty, "The Climate of History: Four Theses," 24.
14. Thompson, "Notes on Exterminism," 31.
15. Marx and Engels, "Manifesto of the Communist Party," *MECW*, vol. 6, 82.
16. *L'Ordine Nuovo*, May 15, 1919. Quoted in Williams, *Proletarian Order*, 11.

Appendix: Confusions and Misconceptions

1. Bookchin, "Social Ecology versus Deep Ecology."
2. Quoted in Magdoff et al., *What Every Environmentalist Needs to Know*, 32.
3. Suckling, "Against the Anthropocene."
4. Moore, *Capitalism in the Web of Life*, 169–73.
5. Baskin, "The Ideology of the Anthropocene."
6. See http://petitions.moveon.org/sign/against-the-official-2.
7. Crist, "On the Poverty of Our Nomenclature," 129, 130, 133, 140, 141.
8. Purdy, "Anthropocene Fever."
9. Crutzen, "The Geology of Mankind," 23.
10. Steffen et al., *Global Change and the Earth System*, 89–90.
11. Ibid., 96.
12. Ibid., 102.
13. Ibid., 140.

14. Ibid., 305.
15. Steffen et al., "The Anthropocene: From Global Change to Planetary Stewardship," 746.
16. Steffen et al., "Planetary Boundaries: Guiding Human Development," 9.
17. Steffen et al., "Trajectory of the Anthropocene," 91.

BIBLIOGRAPHY

Abramsky, Kolya. "Racing to 'Save' the Economy and the Planet: Capitalist or Post-Capitalist Transition to a Post-Petrol World." In *Sparking a Worldwide Energy Revolution*, edited by Kolya Abramsky. Oakland, CA: AK Press, 2010, 5–30.

Adams, Jonathan, Mark Maslin, and Ellen Thomas. "Sudden Climate Transitions during the Quaternary." *Progress in Physical Geography* 23/1 (1999): 1–36.

"All in Good Time." *Nature*, March 12, 2015, 129–30.

Alley, Richard B. *Abrupt Climate Change: Inevitable Surprises*. Washington, DC: National Academy Press, 2002.

———. *The Two-Mile Time Machine: Ice Cores, Abrupt Climate Change, and Our Future*. Princeton: Princeton University Press, 2000.

Alverson, K., I. Laroque, and C. Krull. "Appendix A: The Past Global Changes (PAGES) Program." In *Paleoclimate, Global Change and the Future*, edited by Keith D. Alverson, Raymond S. Bradley, and Thomas F. Pedersen. Berlin: Springer, 2003, 169–73.

Amsterdam Declaration on Global Change, http://www.colorado.edu/AmStudies/lewis/ecology/gaiadeclar.pdf.

Andersen, Stephen, and K. Madhava Sarma. *Protecting the Ozone Layer: The United Nations History*. New York: Routledge, 2002.

Anderson, Kevin, and Alice Bows. "A 2°C Target? Get Real, Because 4°C Is on Its Way." *Parliamentary Brief*, 2010, 19.

———. "Beyond 'Dangerous' Climate Change: Emission Scenarios for a New World." *Philosophical Transactions of the Royal Society* 369 (2011): 20–44.

Angus, Ian, and Simon Butler. *Too Many People?: Population, Immigration, and the Environmental Crisis*. Chicago: Haymarket Books, 2011.

Angus, Ian. "Hijacking the Anthropocene." *Climate & Capitalism,* May 19, 2015, http://climateandcapitalism.com/2015/05/19/hijacking-the-anthropocene/.

————. *The Global Fight for Climate Justice: Anticapitalist Responses to Global Warming and Environmental Destruction*. Black Point, Nova Scotia: Fernwood, 2010.

Archer, David. *The Long Thaw: How Humans Are Changing the Next 100,000 Years of Earth's Climate*. Princeton: Princeton University Press, 2009.

Bahro, Rudolf. *Avoiding Social and Ecological Disaster*. Bath: Gateway Books, 1994.

Baran, Paul A., and Paul M. Sweezy. *Monopoly Capital: An Essay on the American Economic and Social Order*. New York: Monthly Review Press, 1966.

Barnosky, Anthony D. "Did the Anthropocene Begin with a Bang or a Drumroll?" *Huffington Post,* January 20, 2015, http://www.huffingtonpost.com/anthony-d-barnosky/did-the-anthropocene-begin-with-a-bang_b_6494076.html.

————. *Dodging Extinction: Power, Food, Money and the Future of Life on Earth*. Oakland: University of California Press, 2014.

Barnosky, Anthony D., Elizabeth A. Hadly, Jordi Bascompte, Eric L. Berlow, James H. Brown, Mikael Fortelius, Wayne M. Getz et al. "Approaching a State Shift in Earth's Biosphere." *Nature* 486 (June 7, 2012): 52–58.

Baskin, Jeremy. "The Ideology of the Anthropocene?" Melbourne Sustainable Society Institute, 2014, http://sustainable-dev.unimelb.edu.au/sites/default/files/docs/MSSI-ResearchPaper-3_Baskin_2014.pdf.

Beckman, Theodore N. "A Brief History of the Gasoline Service Station." *Journal of Historical Research in Marketing* 3/2 (2011): 156–72.

Biel, Robert. *The Entropy of Capitalism*. Leiden: Brill, 2012.

Bookchin, Murray [Lewis Herber]. *Our Synthetic Environment*. New York: Knopf, 1962.

————. "Social Ecology versus Deep Ecology: A Challenge for the Ecology Movement." http://dwardmac.pitzer.edu/Anarchist_Archives/bookchin/socecovdeepeco.html.

Braverman, Harry. *Labor and Monopoly Capital: The Degradation of Work in the Twentieth Century*. 25th Anniversary ed. New York: Monthly Review Press, 1998.

Brecht, Bertolt. *Tales from the Calendar*. London: Methuen, 1961.

————. *Brecht on Theatre*. New York: Hill and Wang, 1964.

Broeker, Wallace. "Cooling for the Tropics." *Nature* 376 (1995): 212–13.

Brodine, Virginia. *Green Shoots, Red Roots.* New York: International
    Publishers, 2007.

Brown, Lester. "Pavement Is Replacing the World's Croplands." *Grist,*
    March 1, 2001, http://grist.org/article/rice/.

Budyko, M.I. "Polar Ice and Climate." In *Soviet Data on the Arctic Heat
    Budget and Its Climatic Influence,* edited by J. O. Fletcher, B.
    Keller, and S. M. Olenicoff, 9–23. Santa Monica, CA: Rand
    Corporation, 1966.

———. "The Effect of Solar Radiation on the Climate of the Earth." *Tel-
    lus* 21/5 (October 1969); 611–14.

———. "Comments." *Journal of Applied Meteorology* 9 (April 1970): 310.

———. *The Evolution of the Biosphere.* Boston: D. Reidel Publishing, 1986.

Burkett, Paul. *Marx and Nature: A Red and Green Perspective.* New York:
    St. Martin's Press, 1999.

———. *Marxism and Ecological Economics: Toward a Red and Green Po-
    litical Economy.* Leiden: Brill, 2006.

Burroughs, William James. *Climate Change in Prehistory: The End of the
    Reign of Chaos.* Cambridge: Cambridge University Press, 2005.

Buxton, Nick, and Ben Hayes. "Introduction: Security for Whom in a
    Time of Climate Crisis?" In *The Secure and the Dispossessed,*
    edited by Nick Buxton and Ben Hayes, 1–19. London: Pluto
    Press, 2016.

Cagin, Seth, and Philip Dray. *Between Earth and Sky: How CFCs
    Changed Our World and Endangered the Ozone Layer.* New
    York: Pantheon, 1993.

Carson, Rachel. *Lost Woods.* Boston: Beacon Press, 1998.

———. *Silent Spring.* Fortieth Anniversary ed. New York: Houghton
    Mifflin, 2002.

Carty, Tracy, and John Magrath. *Growing Disruption: Climate Change,
    Food, and the Fight Against Hunger.* London: Oxfam, 2013.

Chandler, Alfred D., and Takashi Hikino. *Scale and Scope: The Dynam-
    ics of Industrial Capitalism.* Cambridge, MA: Harvard Univer-
    sity Press, 1990.

Chicago Tribune. "Rockefeller Profits from the Marshall Plan." *Chicago
    Tribune,* December 13, 1948, 16.

Clapp, Jennifer. *Food.* Cambridge: Polity Press, 2012.

Clark, Brett, and Richard York. "Carbon Metabolism: Global Capital-
    ism, Climate Change, and the Biospheric Rift." *Theory and So-
    ciety* 34 (2005): 391–428.

Climate Action Tracker. "2.7°C Is Not Enough—We Can Get Lower."

Update, December 8, 2015. http://climateactiontracker.org/assets/publications/briefing_papers/CAT_Temp_Update_COP21.pdf.

Coady, David, Ian Perry, Louis Sears, and Baoping Shang. *How Large Are Global Energy Subsidies?* Washington, DC: International Monetary Fund, 2015.

Cochran, Bert. *Labor and Communism: The Conflict that Shaped American Unions.* Princeton: Princeton University Press, 1977.

Cohen, Lizabeth. *A Consumers' Republic: The Politics of Mass Consumption in Postwar America.* New York: Vintage Books, 2004.

Commoner, Barry. "Oil, Energy and Capitalism: An Unpublished Talk." *Climate & Capitalism,* July 30, 2013, http://climateandcapitalism.com/2013/07/30/exclusive-an-unpublished-talk-by-barry-commoner/.

———. "Threats to the Integrity of the Nitrogen Cycle: Nitrogen Compounds in Soil, Water, Atmosphere and Precipitation." In *Global Effects of Environmental Pollution,* edited by S. Fred Singer. New York: Springer, 1970.

———. *Making Peace with the Planet.* New York: Pantheon, 1990.

———. *Science and Survival.* New York: Viking , 1966.

———. *The Closing Circle: Nature, Man, and Technology.* New York: Knopf, 1971.

———. *The Poverty of Power: Energy and the Economic Crisis.* New York: Knopf, 1976.

Costa, Alexandre. "Socialism Is Not Possible on a Ruined Planet." *Climate & Capitalism*, April 17, 2014. http://climateandcapitalism.com/2014/04/17/socialism-possible-ruined-planet/.

Coumou, Dim, and Stefan Rahmstorf. "A Decade of Weather Extremes." *Nature Climate Change* 2 (2012): 491–96.

Cowan, Ruth Schwartz. "The Industrial Revolution in the Home." In *The Social Shaping of Technology: How the Refrigerator Got Its Hum,* edited by Donald A. MacKenzie and Judy Wajcman, 181–201. Philadelphia: Open University Press, 1985.

Crist, Eileen. "On the Poverty of Our Nomenclature." *Environmental Humanities* 3 (2013): 129–47.

Crutzen, Paul J. "Geology of Mankind." *Nature* 415/3 (January 3, 2002): 23.

———. "My Life with O_3, NO_x and Other YZO$_x$s." Nobel Lecture. http://www.nobelprize.org/nobel_prizes/chemistry/laureates/1995/crutzen-lecture.pdf.

Crutzen, Paul J., and Eugene F. Stoermer. "The Anthropocene." *Global Change Newsletter*, May 1, 2000, 17.

Crutzen, Paul J., and Will Steffen. "How Long Have We Been in the Anthropocene Era?" *Climatic Change* 61/3 (2003); 251–57.

Custers, Peter. *Questioning Globalized Militarism: Nuclear and Military Production and Critical Economic Theory*. Monmouth, Wales: Merlin Press, 2007.

Davis, Mike. "Who Will Build the Ark?" *New Left Review* 61 (Jan–Feb 2010): 29–46.

———. *Prisoners of the American Dream*. London: Verso, 2000.

Davis, Mike, and Daniel Bertrand Monk, eds. *Evil Paradises: Dreamworlds of Neoliberalism*. New York: New Press, 2007.

Delucchi, Mark A., and Mark Z. Jacobson. "Providing All Global Energy with Wind, Water, and Solar Power, Part II: Reliability, System and Transmission Costs, and Policies." *Energy Policy* 39 (2011): 1170–90.

Diffenbaugh, Noah S., and Martin Scherer. "Observational and Model Evidence of Global Emergence of Permanent, Unprecedented Heat in the 20th and 21st Centuries." *Climatic Change* 107 (2011): 615–24.

Dimitri, Carolyn, Anne Effland, and Neilson Conklin. "The 20th-Century Transformation of U.S. Agriculture and Farm Policy: Economic Information Bulletin Number 3." Washington, DC: U.S. Department of Agriculture, June 1, 2005.

Dimuzio, Tim. "Capitalizing a Future Unsustainable: Finance, Energy and the Fate of Market Civilization." *Review of International Political Economy* 19/3 (2011): 363–88.

Dowd, Douglas. *Capitalism and Its Economics: A Critical History*. London: Pluto Press, 2004.

———. *Inequality and the Global Economic Crisis*. London: Pluto Press, 2009.

Drucker, Peter. "What to Do about Strikes." *Collier's Weekly*, January 18, 1947, 13, 26–27.

DuBoff, Richard B. *Accumulation and Power: An Economic History of the United States*. Armonk, NY: M. E. Sharpe, 1989.

Dunne, John P., Ronald J. Stouffer, and Jasmin G. John. "Reductions in Labour Capacity from Heat Stress under Climate Warming." *Nature Climate Change* 3 (2013): 563–66.

Eckersley, Robyn. "Environmental Security, Climate Change, and Globalizing Terrorism." In *Rethinking Insecurity, War and Violence:*

Beyond Savage Globalization, edited by Damian Grenfell and Paul James, 85–97. London: Routledge, 2009.

Edburg, Rolf, and Alexei Yablokov, *Tomorrow Will Be Too Late*. Tucson: University of Arizona Press, 1991.

Eyring, V., H. W. Kohler, A. Lauer, and B. Lemper. "Emissions from International Shipping: 2. Impact of Future Technologies on Scenarios until 2050." *Journal of Geophysical Research* 110/D17 (2005): 1–18.

Falkowski, P., R. J. Scholes, E. Boyle, J. Canadell, D. Canfield, J. Elser, N. Gruber et al. "The Global Carbon Cycle: A Test of Our Knowledge of Earth as a System." *Science* 290/5490 (October 13, 2000): 291–96.

Fedorov, E. *Man and Nature*. New York: International Publishers, 1972.

Field, Christopher B., Vicente Barros, Thomas F. Stocker, Qin Dahe, David Jon Dokken, Gian-Kasper Plattner, Kristie L. Ebi et al. *Managing the Risks of Extreme Events and Disasters to Advance Climate Change Adaptation: Special Report of the Intergovernmental Panel on Climate Change*. Cambridge: Cambridge University Press, 2012.

Foner, Philip S. *The Great Labor Uprising of 1877*. New York: Monad Press, 1977.

Foster, John Bellamy. "The Great Capitalist Climacteric: Marxism and 'System Change Not Climate Change.'" *Monthly Review* (November 2015): 1–18.

———. "Late Soviet Ecology and the Planetary Crisis." *Monthly Review* (June 2015): 1–20.

———. *Marx's Ecology: Materialism and Nature*. New York: Monthly Review Press, 2000.

———. *The Ecological Revolution: Making Peace with the Planet*. New York: Monthly Review Press, 2009.

———. *The Theory of Monopoly Capitalism: An Elaboration of Marxian Political Economy*. New York: Monthly Review Press, 1986.

———. *The Vulnerable Planet: A Short Economic History of the Environment*. 2nd ed. New York: Monthly Review Press, 1999.

Foster, John Bellamy, and Brett Clark. *The Ecological Rift: Capitalism's War on the Earth*. New York: Monthly Review Press, 2010.

Foster, John Bellamy, Hannah Holleman, and Robert W. McChesney. "The U.S. Imperial Triangle and Military Spending." *Monthly Review* (October 2008): 1–19.

Foster, John Bellamy, and Robert W. McChesney. *The Endless Crisis: How*

Monopoly-Finance Capital Produces Stagnation and Upheaval from the U.S.A. to China. New York: Monthly Review Press, 2012.

Fressoz, Jean-Baptiste. "Losing the Earth Knowingly: Six Environmental Grammars around 1800." In *The Anthropocene and the Global Environmental Crisis: Rethinking Modernity in a New Epoch,* edited by Clive Hamilton, Christophe Bonneuil, and François Gemenne, 70–83. New York: Routledge, 2015.

Fuentes-Nieva, Ricardo, and Nick Galasso. "Working for the Few: Political Capture and Economic Inequality." Oxfam International Briefing Paper, January 2014.

Galbraith, John Kenneth. *The New Industrial State.* Princeton: Princeton University Press, 2007.

Gallopin, Gilberto, Al Hammond, Paul Raskin, and Rob Swart. *Branch Points: Global Scenarios and Human Choice.* Stockholm: Stockholm Environment Institute, 1997.

Ganopolski, Andrey, and Stefan Rahmstorf. "Rapid Changes of Glacial Climate Simulated in a Coupled Climate Model." *Nature* 409 (2001): 153–58.

Gleick, Peter H. "Water, Drought, Climate Change, and Conflict in Syria." *Journal of the American Meteorological Society* (July 2014): 331–40.

Global Humanitarian Forum. *Human Impact Report: Climate Change— The Anatomy of a Silent Crisis.* Geneva: Global Humanitarian Forum, 2009.

Goff, Stan. "Exterminism and the World in the Wake of Katrina." *From the Wilderness,* http://www.fromthewilderness.com/free/ww3/102305_exterminism_katrina.shtml.

Golley, Frank Benjamin. *A History of the Ecosystem Concept in Ecology.* New Haven: Yale University Press, 1993.

GRAIN. "The Exxons of Agriculture." September 30, 2015, https://www.grain.org/article/entries/5270-the-exxons-of-agriculture.pdf.

Haldane, J. B. S. "The Origin of Life." In J. D. Bernal, *The Origin of Life.* New York: World Publishing, 1967, 242-49.

Hamilton, Clive. "Human Destiny in the Anthropocene." In *The Anthropocene and the Global Environmental Crisis,* edited by Clive Hamilton, François Gemenne, and Christophe Bonneuil. 32–43. New York: Routledge, 2015.

———. "Can Humans Survive the Anthropocene?," http://clivehamilton.com/wp-content/uploads/2014/05/Can-humans-survive-the-Anthropocene.pdf.

———. "Ecologists Butt Out: You Are Not Entitled to Redefine the Anthropocene." August 11, 2014, http://clivehamilton.com/ecologists-butt-out-you-are-not-entitled-to-redefine-the-anthropocene/.

———. "Getting the Anthropocene So Wrong." *Anthropocene Review* (2015): 102–7.

———. "The Anthropocene: Too Serious for Post-Modern Games." August 19, 2014, http://clivehamilton.com/the-anthropocene-too-serious-for-post-modern-games.

———. "The New Environmentalism Will Lead Us to Disaster." *Scientific American*, June 19, 2014, http://www.scientificamerican.com/article/the-new-environmentalism-will-lead-us-to-disaster/.

———. "The Theodicy of the 'Good Anthropocene.'" June 1, 2015, http://clivehamilton.com/the-theodicy-of-the-good-anthropocene/.

Hamilton, Clive, François Gemenne, and Christophe Bonneuil. "Thinking the Anthropocene." In *The Anthropocene and the Global Environmental Crisis*, edited by Clive Hamilton, François Gemenne, and Christophe Bonneuil, 1–13. New York: Routledge, 2015.

Hamilton, Clive, and Jacques Grinevald. "Was the Anthropocene Anticipated?" *Anthropocene Review* (2015): 59–72.

Hansen, James, M. Sato, P. Hearty, R. Ruedy, M. Kelley, V. Masson-Delmotte, G. Russell et al. "Ice Melt, Sea Level Rise and Superstorms: Evidence from Paleoclimate Data, Climate Modeling, and Modern Observations that 2°C Global Warming Is Highly Dangerous." *Atmospheric Chemistry and Physics*. July 23, 2015. http://www.atmos-chem-phys-discuss.net/15/20059/2015/acpd-15-20059-2015.pdf.

Hansen, James, Makiko Sato, and Reto Ruedy. "Perception of Climate Change." *PNAS* (2012): E2415–33.

Hansen, James. *Storms of My Grandchildren: The Truth about the Coming Climate Catastrophe and Our Last Chance to Save Humanity*. London: Bloomsbury, 2009.

Harman, Chris. *A People's History of the World: From the Stone Age to the New Millennium*. London: Verso, 2008.

———. *Zombie Capitalism: Global Crisis and the Relevance of Marx*. London: Bookmarks Publications, 2009.

Harnecker, Marta. *Rebuilding the Left*. London: Zed Books, 2007.

Harrington, Michael. *The Other America: Poverty in the United States*. New York: Simon and Schuster, 1993 [1962].

Harrould-Kolieb, Ellycia. "Shipping Impacts on Climate: A Source with Solutions." *Oceana*, July 1, 2008, http://oceana.org/sites/default/files/reports/Oceana_Shipping_Report1.pdf.

Hartman, Chester W., and Gregory D. Squires. *There Is No Such Thing as a Natural Disaster: Race, Class, and Hurricane Katrina.* New York: Taylor & Francis, 2006.

Harvey, David. *The Enigma of Capital and the Crises of Capitalism.* London: Profile Books, 2010.

———. *Seventeen Contradictions and the End of Capitalism.* Oxford: Oxford University Press, 2014.

Hayes, Ben. "Colonizing the Future: Climate Change and International Security Strategies." In *The Secure and the Dispossessed*, edited by Nick Buxton and Ben Hayes, 39–62. London: Pluto Press, 2016.

Heartfield, James. *Unpatriotic History of the Second World War.* London: Zero Books, 2012.

Hibbard, Kathy A., Paul J. Crutzen, Eric F. Lambin, Diana M. Liverman, Nathan J. Mantua, John R. McNeill, Bruno Messerli, and Will Steffen. "Group Report: Decadal-Scale Interactions of Humans and the Environment." In *Sustainability or Collapse? An Integrated History and Future of People on Earth.* Cambridge, MA: MIT Press, 2007.

Hobsbawm, E. J. *The Age of Extremes: The Short Twentieth Century, 1914–1991.* London: Abacus Books, 1995.

Hodges, Dan. "Drown an Immigrant to Save an Immigrant: Why Is the Government Borrowing Policy from the BNP?" *The Telegraph*, October 24, 2014.

Hoppe, Robert A. "Structure and Finances of U.S. Farms: Economic Information Bulletin Number 132." Washington, DC: United States Department of Agriculture, 2014.

Huber, Matthew T. *Lifeblood: Oil, Freedom, and the Forces of Capital.* Minneapolis: University of Minnesota Press, 2013.

Hutchinson, G. Evelyn. "The Biosphere." *Scientific American* 233/3 (1970): 45–53.

Hynes, Patricia. "Pentagon Pollution." *Climate & Capitalism*, February 8, 2005, http://climateandcapitalism.com/2015/02/08/pentagon-pollution-1-war-true-tragedy-commons/.

Ibrahim, Fawzi. *Capitalism versus Planet Earth: An Irreconcilable Conflict.* London: Muswell Press, 2012.

"ICSU Sets Up IGBP Committee." *Eos* 68/11 (1987).

IPCC. "Summary for Policymakers." In *Climate Change 2014: Mitigation of Climate Change*. Cambridge: Cambridge University Press, 2014.

IPCC. *Climate Change 2014: Impacts, Adaptation, and Vulnerability. Part A: Global and Sectoral Aspects*. Cambridge: Cambridge University Press, 2014.

Jacobson, Mark Z., and Mark A. Delucchi. "Providing All Global Energy with Wind, Water, and Solar Power, Part I: Technologies, Energy Resources, Quantities and Areas of Infrastructure, and Materials." *Energy Policy* 39 (2011): 1154–69.

Jacobson, Mark Z., Mark A. Delucchi, Guillaume Bazouin, Zack A. F. Bauer, Christa C. Heavey, Emma Fisher, Sean B. Morris et al. "100% Clean and Renewable Wind, Water, and Sunlight (WWS) All-sector Energy Roadmaps for the 50 United States." *Energy & Environmental Science* 8/7 (2015): 2093–2117.

Jansson, Bruce S. *The Sixteen-Trillion-Dollar Mistake: How the U.S. Bungled Its National Priorities from the New Deal to the Present*. New York: Columbia University Press, 2001.

Karl, Thomas R., and Richard W. Katz. "A New Face for Climate Dice." *Proceedings of the National Academy of Sciences* 109/37 (2012): 14720–21.

Kennedy, David M. *Freedom from Fear: The American People in Depression and War, 1929–1945*. New York: Oxford University Press, 1999.

Klare, Michael T. *Blood and Oil: The Dangers and Consequences of America's Growing Dependency on Imported Petroleum*. New York: Henry Holt, 2004.

Klein, Naomi. *This Changes Everything: Capitalism vs. the Climate*. Toronto: Knopf, 2014.

Klinenberg, Eric. *Heat Wave: A Social Autopsy of Disaster in Chicago*, 2nd ed. Chicago: University of Chicago Press, 2015.

Kumar, Sanjay. "Green Climate Fund Faces Slew of Criticism." *Nature* 527/7579 (November 20, 2015): 419–20.

Kump, Lee R. "The Last Great Global Warming." *Scientific American*, July 1, 2011.

Kunstler, James Howard. *The Long Emergency: Surviving the Converging Catastrophes of the Twenty-First Century*. New York: Atlantic Monthly Press, 2005.

Kunzig, Robert. "Hothouse Earth." *National Geographic*, October 1, 2011.

Lancet. "Managing the Health Effects of Climate Change." *The Lancet* 373 (2009): 1693–1733.

———. "Climate Change and Health: On the Latest IPCC Report." *The Lancet* 383 (2014): 187.

———. "Health and Climate Change: Policy Responses to Protect Public Health." *The Lancet* 386 (2015): 1–53.

Lebowitz, Michael A. *The Socialist Alternative: Real Human Development.* New York: Monthly Review Press, 2010.

Lenin, V. I. "What Is to Be Done?" In *Collected Works*, vol. 5. Moscow: Progress Publishers, 1964, 347–530.

———. "Imperialism, the Highest Stage of Capitalism." In *Collected Works*, vol. 22. Moscow: Progress Publishers, 1964, 185–304.

Leopold, Aldo, and Charles Walsh Schwartz. *A Sand County Almanac with Other Essays on Conservation from Round River (1949).* New York: Ballantine Books, 1970.

Levins, Richard. "Why Programs Fail." *Monthly Review* 61/10 (2010): 43–49.

Levins, Richard, and Richard Lewontin. *The Dialectical Biologist.* Cambridge, Massachusetts: Harvard University Press, 1985.

Lewontin, Richard. "Agricultural Research and the Penetration of Capital." *Science for the People*, 1982, 12–17.

Linden, Eugene. "Storm Warnings Ahead." *Time*, April 5, 2004. 58.

Lipsitz, George. *Rainbow at Midnight: Labor and Culture in the 1940s.* Chicago: University of Illinois Press, 1994.

Longo, Stefano, Rebecca Clausen, and Brett Clark. *The Tragedy of the Commodity: Oceans, Fisheries, and Aquaculture.* New Brunswick, NJ: Rutgers University Press, 2015.

Lott, Fraser C., Nikolaos Christidis, and Peter A. Stott, "Can the 2011 East African Drought Be Attributed to Human-Induced Climate Change?" *Geophysical Research Letters* 40/6 (March 2013): 1177–81.

Lovelock, J. E., R. J. Maggs, and R. J. Wade. "Halogenated Hydrocarbons in and over the Atlantic." *Nature* 241 (1973): 194–96.

Löwy, Michael. *Ecosocialism: A Radical Alternative to Capitalist Catastrophe.* Chicago: Haymarket Books, 2015.

Luxemburg, Rosa. *The Rosa Luxemburg Reader.* Edited by Peter Hudis and Kevin Anderson. New York: Monthly Review Press, 2004.

———. *Selected Political Writings.* Edited by Dick Howard. New York: Monthly Review Press, 1971,

Magdoff, Fred, and John Bellamy Foster. *What Every Environmentalist*

Needs to Know about Capitalism: A Citizen's Guide to Capitalism and the Environment. New York: Monthly Review Press, 2011.

Magdoff, Fred. "Ecological Civilization." *Monthly Review* 62/8 (January, 2011): 1–25.

Malm, Andreas. "China as Chimney of the World: The Fossil Capital Hypothesis." *Organization & Environment* 25/2 (2012): 146–77.

———. "The Origins of Fossil Capital: From Water to Steam in the British Cotton Industry." *Historical Materialism* 21/1 (2013): 15–68.

———. *Fossil Capital: The Rise of Steam Power and the Roots of Global Warming.* London: Verso, 2016.

Mandel, Ernest. *Late Capitalism.* Rev. ed. London: Verso, 1978.

———. *Long Waves of Capitalist Development: A Marxist Interpretation.* Rev. ed. London: Verso, 1995.

———. *Marxist Economic Theory.* Translated by Brian Pearce. New York: Monthly Review Press, 1968.

———. *The Second Slump: A Marxist Analysis of Recession in the Seventies.* London: Verso, 1980.

Marcott, Shaun A., Jeremy D. Shakun, Peter U. Clark, and Alan C. Mix. "A Reconstruction of Regional and Global Temperature for the Past 11,300 Years." *Science* 339 (2013): 1198–1201.

Margulis, Lynn, and Dorion Sagan. *What Is Life?* New York: Simon and Schuster, 1995.

Marsh, George P. *Man and Nature, or, Physical Geography as Modified by Human Action.* New York: Charles Scribner, 1864. Repr. *Man and Nature.* Cambridge, MA: Harvard University Press, 1965.

Martinez, Oswaldo. "We Are Facing Something More than a Mere Financial Crisis." Translated by Richard Fidler. *Socialist Voice,* March 23, 2009, http://www.socialistvoice.ca/?p=375.

Marx, Karl, and Frederick Engels. *Collected Works (MECW).* 50 vols. New York: International Publishers, 1975–2004.

Marx, Karl. *Capital.* Vol. 1. Harmondsworth: Penguin Books, 1976.

———. *Capital.* Vol. 3. Harmondsworth: Penguin Books, 1981.

Maschado, Barry F. *In Search of a Usable Past: The Marshall Plan and Postwar Reconstruction.* Vicksburg, VA: George C. Marshall Foundation, 2007.

Maslin, M. A., and S. L. Lewis. "Anthropocene: Earth System, Geological, Philosophical and Political Paradigm Shifts." *Anthropocene Review* 1 (2015): 108–16.

Masters, Jeff. "PETM: Global Warming, Naturally." *Weather Under-
        ground*, http://www.wunderground.com/climate/PETM.asp.
Maxwell, James, and Forrest Briscoe. "There's Money in the Air: The
        CFC Ban and DuPont's Regulatory Strategy." *Business Strategy
        and the Environment* 6 (1997): 276–86.
McNeill, John Robert. *Something New under the Sun: An Environmental
        History of the Twentieth-century World*. New York: W. W. Nor-
        ton, 2000.
Meeting Report: IGBP: Crown Jewel or Prodigal Son? *Eos 70/50* (1989).
Melillo, Jerry M., Terese Richmond, and Gary W. Yohe, eds. *Highlights
        of Climate Change Impacts in the United States: The Third
        National Climate Assessment*. Washington DC: U.S. Global
        Change Research Program, 2014.
Mészáros, István. *The Power of Ideology*. New York: New York University
        Press, 1989.
———. *Socialism or Barbarism: From the "American Century" to the
        Crossroads*. New York: Monthly Review Press, 2001.
———. *The Challenge and Burden of Historical Time: Socialism in the
        Twenty-first Century*. New York: Monthly Review Press, 2008.
———. *The Necessity of Social Control*. New York: Monthly Review Press, 2015.
Meyer, Robinson. "Al Gore Dreamed Up a Satellite—And It Just Took Its
        First Picture of Earth." *The Atlantic*, July 20, 2015, http://www.
        theatlantic.com/technology/archive/2015/07/our-new-and-
        daily-view-of-the-blue-marble/399011/.
Mill, John Stuart. *Essays on Some Unsettled Questions of Political Econo-
        my*. New York: Cosimo, 2007.
Millennium Ecosystem Assessment. "Living Beyond Our Means: Natu-
        ral Assets and Human Well-Being, Statement of the Board of
        Director." March 1, 2005, http://www.millenniumassessment.
        org/documents/document.429.aspx.pdf.
———. "United Nations Launches Extensive Study of Earth's Ecosys-
        tems." News Release, June 5, 2001, http://www.millenniumas-
        sessment.org/en/Articlee5cc.html.
———. *Ecosystems and Human Well-Being: Synthesis*. Washington, DC:
        Island Press, 2005.
Mitchell, Timothy. *Carbon Democracy: Political Power in the Age of Oil*.
        London: Verso, 2011.
Moore III, Berrien. "Challenges of a Changing Earth." In *Challenges of a
        Changing Earth*, W. Steffen, J. Jäger, D. J. Carson, C. Bradshaw,
        editors. Berlin: Springer, 2002.

Moore, Jason W. *Capitalism in the Web of Life: Ecology and the Accumulation of Capital.* London: Verso, 2015.

Muhammad, Umair. *Confronting Injustice: Social Activism in the Age of Individualism.* Toronto: Umair Muhammed, 2014.

Nadeau, Robert L. *The Environmental Endgame: Mainstrean Economics, Ecological Disaster, and Human Survival.* New Brunswick, NJ: Rutgers University Press, 2006.

National Research Council. *Abrupt Impacts of Climate Change: Anticipating Surprises.* Washington, DC: National Academies Press, 2013.

National Research Council. *Climate Stabilization Targets: Emissions, Concentrations, and Impacts over Decades to Millennia.* Washington, DC: National Academies Press, 2011.

Nikiforuk, Andrew. *The Energy of Slaves: Oil and the New Servitude.* Vancouver, BC: Greystone Books, 2012.

Nordhaus, Ted, Michael Shellenberger, and Jenna Makuno. "Ecomodernism and the Anthropocene: Humanity as a Force for Good." *Breakthrough Journal* 5 (Summer 2015), http://thebreakthrough.org/index.php/journal/issue-5/ecomodernism-and-the-anthropocene.

Oakes, Walter J. "Toward a Permanent War Economy?" *Politics* 1/1 (February 1944): 11–17.

Odum, Howard T. *Environment, Power, and Society for the Twenty-First Century.* New York: Columbia University Press, 2007.

Oldfield, F. "When and How Did the Anthropocene Begin?" *Anthropocene Review* 2/2 (2015): 101.

Oldfield, Frank, and Will Steffen. "The Earth System." In *Global Change and the Earth System: A Planet under Pressure,* edited by Will Steffen et al. Berlin: Springer, 2004.

Oparin, A. I. "The Origin of Life." In J. D. Bernal, *The Origin of Life.* New York: World Publishing, 1967, 199–234.

Oreskes, Naomi. "The Scientific Consensus on Climate Change: How Do We Know We're Not Wrong?" In *Climate Change: What It Means for Us, Our Children, Our Grandchildren.* Edited by Joseph Dimento and Pamela Doughman. Cambridge, MA: MIT Press, 2007.

Painter, David S. "The Marshall Plan and Oil." *Cold War History* 9/2 (2009): 159–75.

Parenti, Christian. *Tropic of Chaos: Climate Change and the New Geography of Violence.* New York: Nation Books, 2011.

———. "The Catastrophic Convergence: Militarism, Neoliberalism and Climate Change." In *The Secure and the Dispossessed*, edited by Nick Buxton and Ben Hayes. 23–38. London: Pluto Press, 2016.

Pearce, Fred. *With Speed and Violence: Why Scientists Fear Tipping Points in Climate Change*. Boston: Beacon Press, 2007.

Perfecto, Ivette, John H. Vandermeer, and Angus Wright. *Nature's Matrix: Linking Agriculture, Conservation and Food Sovereignty*. London: Earthscan, 2009.

Petit, J. R., J. Jouzel, D. Raynaud, N. I. Barkov, J.-M. Barnola, I. Basile, M. Bender et al. "Climate and Atmospheric History of the Past 420,000 Years from the Vostok Ice Core, Antarctica." *Nature* 399 (June 3, 1999): 429–36.

Phillips, Kevin. *Wealth and Democracy: A Political History of the American Rich*. New York: Random House, 2002.

Pobodnik, Bruce. *Global Energy Shifts: Fostering Sustainability in a Turbulent Age*. Philadelphia: Temple University Press, 2006.

Potsdam Institute for Climate Impact Research and Climate Analytics. *Turn Down the Heat: Why a 4° Warmer World Must Be Avoided*. Washington, DC: World Bank, 2012.

———. *Turn Down the Heat: Climate Extremes, Regional Impacts and the Case for Resilience*. Washington, DC: World Bank, 2013.

———. *Turn Down the Heat: Confronting the New Climate Normal*. Washington, DC: World Bank, 2014.

Preis, Art. *Labor's Giant Step: Twenty Years of the CIO*. New York: Pioneer Publishers, 1964.

Purdy, Jedediah. "Anthropocene Fever." *Aeon*, March 31, 2015, http://aeon.co/magazine/science/should-we-be-suspicious-of-the-anthropocene/.

Risbey, James S. "The New Climate Discourse: Alarmist or Alarming?" *Global Environmental Change* 18 (June 2007): 26–37.

Roan, Sharon. *Ozone Crisis: The 15-year Evolution of a Sudden Global Emergency*. New York: John Wiley, 1990.

Rockström, Johan. "Bounding the Planetary Future: Why We Need a Great Transition." *Great Transition Initiative*, April 1, 2015, http://www.greattransition.org/publication/bounding-the-planetary-future-why-we-need-a-great-transition.

Rockström, J., W. Steffen, K. Noone, Å. Persson, F. S. Chapin, III, E. Lambin, T. M. Lenton et al. "Planetary Boundaries: Exploring the Safe Operating Space for Humanity." *Ecology and Society*

14/2 (2009): 32. http://www.ecologyandsociety.org/vol14/iss2/art32/.

Rockström, Johan, and Mattias Klum. *Big World, Small Planet: Abundance within Planetary Boundaries.* Stockholm: Max Strom, 2015.

Rockström, Johan, Will Steffen, Kevin Noone, and Åsa Persson. "A Safe Operating Space for Humanity." *Nature* 461 (September 24, 2009): 472–75.

Roederer, Juan G. "ICSU Gives Green Light to IGBP." *Eos* 67/41 (1986).

Ross, Benjamin, and Steven Amter. *The Polluters: The Making of Our Chemically Altered Environment.* New York: Oxford University Press, 2010.

Rowland, Sherwood. "Nobel Lecture in Chemistry, December 8, 1995." Nobel Prize.org, http://www.nobelprize.org/nobel_prizes/chemistry/laureates/1995/rowland-lecture.pdf.

Ruddiman, William F. "How Did Humans First Alter Global Climate?" *Scientific American* 292/3 (March 2005): 46–53.

———. "The Anthropogenic Greenhouse Era Began Thousands of Years Ago." *Climatic Change* 61 (2003): 261–93.

Ruder, Eric. "How Katrina Unleashed a Storm of Racism." *Socialist Worker*, October 14, 2005.

Sanders, Barry. *The Green Zone: The Environmental Costs of Militarism.* Oakland, CA: AK Press, 2009.

Sassen, Saskia. *Expulsions: Brutality and Complexity in the Global Economy.* Cambridge, MA: Harvard University Press, 2014.

Schellnhuber, H. J. "The Laws of Nature—and the Laws of Civilization." Dinner speech at COP 18. https://www.pik-potsdam.de/members/john/highlights/files/dinner-speech-at-cop18-in-doha.

———. "'Earth System' Analysis and the Second Copernican Revolution." *Nature* 402 (1999): C19–C23.

———. "Discourse: Earth System Analysis: The Scope of the Challenge." In *Earth System Analysis: Integrating Science for Sustainability*, edited by Hans Joachim Schellnhuber and V. Wenzel, 3–195. Berlin: Springer, 1998.

Schwartz, Peter, and Doug Randall. *An Abrupt Climate Change Scenario and Its Implications for United States National Security.* October 2003, http://www.climate.org/PDF/clim_change_scenario.pdf.

Sellars, William D. "A Global Climatic Model Based on the Energy Balance of the Earth Atmosphere System," *Journal of Applied Meteorology* 8 (June 1969): 392–400.

Semova, Dimitrina, Joan Pedro, Luis Luján, Ashley Jackson-Lesti, Ryan
         Stevens, Chris Marten, Kristy Nelson et al. "US Department of
         Defense Is the Worst Polluter on the Planet." Project Censored,
         October 2, 2010.. http://www.projectcensored.org/2-us-depart-
         ment-of-defense-is-the-worst-polluter-on-the-planet/.
Seneviratne, Sonia I., Markus G. Donat, Brigitte Mueller, and Lisa V.
         Alexander. "No Pause in the Increase of Hot Temperature Ex-
         tremes." *Nature Climate Change* 4 (2014): 161–63.
Shah, Sonia. *Crude: The Story of Oil*. New York: Seven Stories Press,
         2011.
Shantser, E. V. "The Anthropogenic System (Period)." In *Great Soviet
         Encyclopedia*, vol. 2. NewYork: Macmillan, 1973, 139–44.
Simms, Andrew. *Ecological Debt: Global Warming and the Wealth of Na-
         tions*. 2nd ed. London: Pluto Press, 2009.
Singer, Daniel. *Whose Millennium? Theirs or Ours?* New York: Monthly
         Review Press. 1999.
Smil, Vaclav. *Energy at the Crossroads: Global Perspectives and Uncer-
         tainties*. Cambridge, MA: MIT Press, 2003.
Solnit, Rebecca. *A Paradise Built in Hell: The Extraordinary Communi-
         ties that Arise in Disaster*. New York: Penguin, 2010.
Speth, James Gustave. *The Bridge at the Edge of the World: Capitalism,
         the Environment, and Crossing from Crisis to Sustainability*.
         New Haven: Yale University Press, 2008.
Spitz, Peter H. *Petrochemicals: The Rise of an Industry*. New York: Wiley,
         1988.
Steffen, Will, "An Integrated Approach to Understanding Earth's Me-
         tabolism." *Global Change Newsletter* 41 (May 2000): 9–10, 16.
———. "Commentary." In *The Future of Nature*, 486–90. New Haven:
         Yale University Press, 2013.
Steffen, Will, and Peter Tyson. *IGBP Science No. 4. Global Change and the
         Earth System: A Planet under Pressure*. Stockholm: IGBP, 2001.
Steffen, Will, A. Sanderson, P. D. Tyson, J. Jager, P. A. Matson, B. Moore
         III, F. Oldfield et al. *Global Change and the Earth System: A
         Planet under Pressure*. Berlin: Springer, 2004.
Steffen, Will, and Mark Stafford Smith. "Planetary Boundaries, Equity
         and Global Sustainability: Why Wealthy Countries Could
         Benefit from More Equity." *Current Opinion in Environmental
         Sustainability* 5 (2013): 403–8.
Steffen, Will, Åsa Persson, Lisa Deutsch, Jan Zalasiewicz, Mark Williams,
         Katherine Richardson, Carole Crumley et al. "The Anthropo-

cene: From Global Change to Planetary Stewardship." *AMBIO: A Journal of the Human Environment* 40 (2011): 739–61. http://link.springer.com/article/10.1007/s13280-011-0185-x.

Steffen, Will, Jacques Grinevald, Paul Crutzen, and John McNeill. "The Anthropocene: Conceptual and Historical Perspectives." *Philosophical Transactions of the Royal Society A* 369 (2011), 842–67.

Steffen, Will, Jill Jäger, David J. Carson, and Clare Bradshaw. *Challenges of a Changing Earth: Proceedings of the Global Change Open Science Conference.* Amsterdam, The Netherlands, 10–13 July 2001. Berlin: Springer, 2002.

Steffen, Will, Katherine Richardson, Johan Rockström, Sarah E. Cornell, Ingo Fetzer, Elena M. Bennett et al. "Planetary Boundaries: Guiding Human Development on a Changing Planet." *Science* 347/6223 (February 13, 2015): 736–47.

Steffen, Will, Meinrat O. Andreae, Bert Bolin, Peter M. Cox, Paul J. Crutzen, Ulrich Cubasch, Hermann Held et al. "Abrupt Changes: The Achilles' Heels of the Earth System." *Environment: Science and Policy for Sustainable Development* 46/3 (2004): 8–20.

Steffen, Will, Paul J. Crutzen, and John R. McNeill. "The Anthropocene: Are Humans Now Overwhelming the Great Forces of Nature?" *AMBIO: A Journal of the Human Environment* 38/8 (2011): 614–21.

Steffen, Will, Wendy Broadgate, Lisa Deutsch, Owen Gaffney, and Cornelia Ludwig. "The Trajectory of the Anthropocene: The Great Acceleration." *Anthropocene Review* 2/ 1 (April, 2015): 81–98.

Suckling, Kieran. "Against the Anthropocene." *Immanence*, July 7, 2014. http://blog.uvm.edu/aivakhiv/2014/07/07/against-the-anthropocene/.

Suvin, Darko. *In Leviathan's Belly: Essays for a Counter-Revolutionary Time.* Rockville, MD: Wildside Press, 2013. Kindle Edition.

Sweezy, Paul M. "Capitalism and the Environment." *Monthly Review* 41/2 (June 1989): 1–10.

Syvitski, James P. M. , Charles J. Vörösmarty, Albert J. Kettner, and Pamela Green. "Impact of Humans on the Flux of Terrestrial Sediment to the Global Coastal Ocean." *Science* 308/5720 (April 15, 2005): 376–80.

Tansley, Arthur. "The Use and Abuse of Vegetational Terms and Concepts." *Ecology* 16/3 (1935): 284–307.

Thompson, E. P. "Notes on Exterminism, the Last Stage of Civilization." *New Left Review* 121 (May–June 1980): 3–31.

———. *Beyond the Cold War.* New York: Pantheon, 1982.

Thorne, J. "Profiteering in the Second World War." Marxists Internet Archive. https://www.marxists.org/history/etol/newspape/fi/vol07/no06/thorne.htm.

Townsend, Terry [Norm Dixon]. "Change the System—Not the Climate!" *Green Left Weekly,* January 26, 2007. https://www.greenleft.org.au/node/36888.

Turgeon, Lynn. *Bastard Keynesianism: The Evolution of Economic Thinking and Policymaking since World War II.* Westport, CT.: Greenwood Press, 1996.

UNFCCC Secretariat. *Report on the Structured Expert Dialogue on the 2013–2015 Review.* Geneva: United Nations Office, 2015.

UNFCCC. "Paris Agreement, December 12, 2015." http://unfccc.int/resource/docs/2015/cop21/eng/l09.pdf.

United Nations Development Program. *Human Development Report, 2007/2008: Fighting Climate Change, Human Solidarity in a Divided World.* New York: Palgrave Macmillan, 2007.

———. *Human Development Report 2011, Sustainability and Equity: A Better Future for All.* New York: Palgrave Macmillan, 2011.

———. *Human Development Report 2013, The Rise of the South: Human Progress in a Diverse World.* New York: UNDP, 2013.

United Nations. *World Economic and Social Survey 2011: The Great Green Technological Transformation.* New York: UN Department of Economic and Social Affairs, 2011.

Vaughan, Adam. "Human Impact Has Pushed Earth into the Anthropocene, Scientists Say." *The Guardian,* January 9, 2016, http://www.theguardian.com/environment/2016/jan/07/human-impact-has-pushed-earth-into-the-anthropocene-scientists-say.

Vernadsky, Vladimir I. *The Biosphere.* New York: Springer-Verlag, 1998.

———. "Some Words about the Noösphere." *150 Years of Vernadsky,* vol. 2: *The Noösphere,* edited by Jason Ross. Washington, DC: 21st Century Science Associates, 2014, 79–84.

Vidal, John. "Shipping Boom Fuels Rising Tide of Global CO_2 Emissions." *The Guardian,* February 13, 2008.

Wagner, Gernot, and Martin L. Weitzman. *Climate Shock: The Economic Consequences of a Hotter Planet.* Princeton: Princeton University Press, 2015.

Waters, C. N., Jan Zalasiewicz, M. Williams, M. A. Ellis, and A. M. Snelling, eds. *A Stratigraphical Basis for the Anthropocene.* London: Geological Society, 2014.

Waters, Colin N., Jan Zalasiewicz, Colin Summerhayes, Anthony D. Barnosky, Clément Poirier, Agnieszka Gałuszka, Alejandro Cearreta et al. "The Anthropocene Is Functionally and Stratigraphically Distinct from the Holocene." *Science* 351/6269 (January 8, 2016): 137, 2622-1–2622-10.

Wayne, Graham. *The Beginner's Guide to Representative Concentration Pathways*. Skeptical Science, 2013. https://www.skepticalscience.com/rcp.php.

Weart, Spencer. "Interview with M. I. Budyko: Oral History Transcript," March 25, 1990, https://www.aip.org/history-programs/niels-bohr-library/oral-histories/31675.

———. *The Discovery of Global Warming*. Cambridge, MA: Harvard University Press, 2003.

Weston, Del. *The Political Economy of Global Warming: The Terminal Crisis*. New York: Routledge, 2014.

White House. The G-20 Toronto Summit Declaration. https://www.whitehouse.gov/the-press-office/g-20-toronto-summit-declaration.

———. *A National Security Strategy of Engagement and Enlargement*. February 1995. http://www.dtic.mil/doctrine/doctrine/research/nss.pdf.

Williams, A. Park, Richard Seager, John T. Abatzoglou, Benjamin I. Cook, Jason E. Smerdon, and Edward R. Cook. "Contribution of Anthropogenic Warming to California Drought during 2012–2014." *Geophysical Research Letters* 42 (2015): 6819–28.

Williams, Chris. *Ecology and Socialism: Solutions to the Capitalist Ecological Crisis*. Chicago: Haymarket Books, 2010.

Williams, Gwyn A. *Proletarian Order: Antonio Gramsci, Factory Councils and the Origins of Italian Communism, 1911–1921*. London: Pluto Press, 1975.

Winebrake, J. J., J. J. Corbett, E. H. Green, A. Lauer, and V. Eyring. "Mitigating the Health Impacts of Pollution from Oceangoing Shipping: An Assessment of Low-Sulfur Fuel Mandates." *Environmental Science & Technology* 43/13 (June 2009): 4776–82.

Wolmar, Christian. *Fire & Steam: How the Railways Transformed Britain*. London: Atlantic Books, 2008.

World Bank. *Building Resilience: Integrating Climate and Disaster Risk into Development. Lessons from World Bank Group Experience*. Washington, DC: World Bank, 2013.

World Economic Forum. *The New Plastics Economy: Rethinking the Future of Plastics.* January 2016. http://www3.weforum.org/docs/WEF_The_New_Plastics_Economy.pdf.

World Health Organization. "Climate Change and Health: WHO Fact Sheet No. 266," August 1, 2014, http://www.who.int/mediacentre/factsheets/fs266/en/.

Wright, Christopher, and Daniel Nyberg. *Climate Change, Capitalism and Corporations: Processes of Creative Self-Destruction.* Cambridge: Cambridge University Press, 2015.

Yates, Michael. *Naming the System: Inequality and Work in the Global Economy.* New York: Monthly Review Press, 2003.

Yergin, Daniel. *The Prize: The Epic Quest for Oil, Money, and Power.* New York: Simon & Schuster, 1991.

Zalasiewicz, Jan, Alan Smith, Tiffany Barry, and Angela L. Coe. "Are We Now Living in the Anthropocene?" *GSA Today* 18/2 (2008): 4–8.

Zalasiewicz, Jan, and Mark Williams. *The Goldilocks Planet: The Four Billion Year Story of Earth's Climate.* Oxford: Oxford University Press, 2012.

Zalasiewicz, Jan, Colin N. Waters, Mark Williams, Anthony D. Barnosky, Alejandro Cearreta, Paul Crutzen, Erle Ellis et al. "When Did the Anthropocene Begin? A Mid-Twentieth-Century Boundary Level Is Stratigraphically Optimal." *Quaternary International*, 383 (October 2015): 196–203.

Zalasiewicz, Jan, M. Williams, A. Haywood, and M. Ellis. "The Anthropocene: A New Epoch of Geological Time?" *Philosophical Transactions of the Royal Society A: Mathematical, Physical and Engineering Sciences* 369/1938 (March 13, 2011): 835–41.

INDEX

Stimson, Henry, 138–39

Stockholm Resilience Center, 71, 73

Stoermer, Eugene, 27, 33, 34–37

Stratigraphic Commission (of Geological Society of London), 50–51

stratigraphy, 50

strikes, 143–44

suburbs, 155–58

Suckling, Keiran, 225

sulfur, 167

Suvin, Darko, 137

Sweezy, Paul M.: on concentration in capitalism, 134–35; on environmental crisis, 78; on innovations, 132; on profits, 117

synthetics, 134

Taft-Hartley Act (1947), 144

Tansley, Arthur, 117

technofossils, 57

temperatures, 56; global averages for, 109; shifts in, 90–103

Thompson, E. P., 15, 179–80, 221

tipping points, 63–66, 104; for agriculture, 102; planetary boundaries distinguished from, 74–76

Townsend, Terry, 204–5

Trotsky, Leon, 186

Truman, Harry S, 142, 144

Turgeon, Lynn, 143

Turn Down the Heat. (World Bank), 97–99

Tutu, Desmond, 184

2100ism, 105

ultraviolet light, 79–80

unemployment, 153

unions, 143–45, 163–64

United Kingdom, *see* Great Britain

United Mine Workers, 144

United Nations, 102–3

United Nations Development Program, 190

United Nations Environment Program, 39

United States: atmospheric carbon dioxide from, 152; automobiles in, 132; climate change as national-security issue for, 182–84; concentration of capital in, 134; Marshall Plan by, 145–46; military spending in, 160–61, 204; permanent war economy in, 141–42; petroleum consumption in, 148; postwar, 138–41; postwar economy in, 142–45; railroads in, 129, 130

Vandermeer, John, 158–59

Vavilov, N. I., 209

Vernadsky, Vladimir I., 11, 13, 209

Vietnam, 163

Villa, Francisco "Pancho," 131

Vostok (Antarctica), 61, 62, 70

Wagner, Gernot, 105

wastes, 167

water power, 129

Waters, Colin, 55

water supply: freshwater use, 74; Millennium Ecosystem Assessment on, 40

Watt, James, 129

Weart, Spencer, 63

Weitzman, Martin, 105